A Brief Tour of
MODERN QUANTUM
MECHANICS

A Brief Tour of
MODERN QUANTUM
MECHANICS

Moshe Gitterman

Bar Ilan University, Israel

World Scientific

NEW JERSEY · LONDON · SINGAPORE · BEIJING · SHANGHAI · HONG KONG · TAIPEI · CHENNAI

Published by

World Scientific Publishing Co. Pte. Ltd.

5 Toh Tuck Link, Singapore 596224

USA office: 27 Warren Street, Suite 401-402, Hackensack, NJ 07601

UK office: 57 Shelton Street, Covent Garden, London WC2H 9HE

British Library Cataloguing-in-Publication Data
A catalogue record for this book is available from the British Library.

A BRIEF TOUR OF MODERN QUANTUM MECHANICS

ISBN-13 978-981-4374-22-4
ISBN-10 981-4374-22-9

Printed in Singapore.

Preface

Anyone who is not shocked by quantum theory, has not understood it. (N. Bohr)

I can safely say that nobody understand quantum mechanics. (R. Feynman)

Many years ago, when I took my first course in quantum mechanics, I was fully confused. I asked many questions, and my professor, a well-known scientist and the author of a quantum mechanics textbook, finally told me, "It is hard to speak with you because you have no imagination". I was very upset by that. How can I become a scientist without imagination. Only later, with a knowledge of the above quotes by N. Bohr and R. Feynman, I realized that I am in a good company. Usually we accept new information using our previous knowledge, but this is not the case for quantum mechanics. Previous knowledge of classical physics makes the understanding of quantum theory even more difficult since one has to reject many ingrained concepts. This is the reason for so many textbooks on quantum mechanics, where each author explains his approach (I am afraid to say, understanding) to quantum mechanics. That is the reason for my decision to add one more book to this long list, hoping that my journey to quantum mechanics and the course of lectures that I taught might be useful for the next generation of scientists.

The main problem in teaching quantum mechanics is the inhomogeneity of the audience, made up of both strong and weak students. A teacher must make sure that the material is both clear for the weak student and interesting for the strong student. Advanced quantum mechanics is intended for those students who have chosen science as their profession and, consequently, their serious attitude to this course is anticipated. My quantum mechanics course was improved as years went by, subject to these circumstances.

My aim was to make the presentation as simple as possible, so that the student with only a general knowledge of elementary quantum mechanics and a limited knowledge of mathematical physics could easily find in this small volume all the required information for his or her own theoretical or experimental work. The organization of the book is as follows.

The introductory Chapter 1 contains a general description of classical mechanics, electrodynamics and elementary quantum mechanics, which are required for understanding of material described in the following chapters. Chapter 2 describes the interaction between a quantum particle and a classical electromagnetic field (transition probabilities between different quantum states, selection rules, etc.), including such key phenomena as the photoelectric effect, Cherenkov effect and Compton effect. Chapter 3 is devoted to the main properties of many-body systems with a comprehensive macroscopic and microscopic description of systems of fermions and bosons, including superfluidity and superconductivity, second quantization and a full quantum description of the interaction between particles and fields. Chapter 4 deals with Feynman diagrams, which permit exact calculations of the S-matrix and Green function by Hartree and Hartee-Fock methods or by approximate analysis (electron gas at high and low densities). Relativistic quantum mechanics is described in Chapter 5, including the Klein-Gordon and Dirac equations, motion of a free electron and a particle in a central field, and the properties of the physical vacuum (Lamb shift, Klein paradox and Casimir effect). Chapters 6 and 7 contain an extended treatment of two phenomena, which are generally briefly described in an undergraduate course of quantum mechanics, namely, the connection between quantum and classical mechanics (WKB approximation) and scattering phenomena. The tunneling of a particle through a potential barrier and its energy levels in a potential well provide the subject matter of Chapter 6, whereas Chapter 7 contains the description of the two main methods of treating scattering theory: the Born approximation and the method of partial waves. Chapter 8 describes qualitative methods in quantum mechanics, which allow one to obtain results which follow from symmetry requirements and the general properties of space and time. Finally, Chapter 9 contains some quantum paradoxes, which provide insight into the strange quantum mechanical laws. Complex mathematical expressions are avoided throughout the book. Some problems with solutions are given at the end of Chapters 2-7. Many more examples can be found in books of problems [1]-[9].

I would like to thank Prof. Nathan Aviezer for his help and useful advice.

Contents

Chapter 1

Introduction

In this slim volume, we will briefly trace the development of physics from the mechanics of Newton and the electrodynamics of Maxwell to the frontiers of modern physics. We shall concentrate on basic principles, showing that knowledge of the basics often suffices to explain many modern phenomena.

1.1   Classical mechanics

1.1.1   Newton's law

All of Newtonian mechanics is encapsulated in the equation, $F = md^2x/dt^2$. Here, F is the force applied to a particle of mass m, situated at position $x(t)$ at time t, which causes the particle to accelerate. The goal of mechanics is to find $x(t)$, the position of the particle at any future time t, given its position and velocity at some initial time t_0.

We shall illustrate many of our formulae by the simple one-dimensional harmonic oscillator, for which the force is $F = -kx$, a force that attempts to return the oscillator to its equilibrium position. Inserting this force into Newton's equation gives

$$m\frac{d^2x}{dt^2} = -kx, \tag{1.1}$$

a differential equation that is easily solved to yield

$$x = C\sin\left(\omega t + \alpha\right) \tag{1.2}$$

where $\omega^2 = k/m$ and the two constants C and α are determined by the initial position and initial velocity of the oscillator.

If the whole world were to consist solely of uncoupled harmonic oscillators, classical mechanics could end right here. However, most mechanical

systems are much more complicated. Therefore, it is convenient to introduce new quantities that simplify the calculation. An additional advantage
of these new quantities is that they pave the way in going from classical
mechanics to quantum mechanics.

First and foremost, for non-dissipative forces (no friction), one introduces the potential energy $U(x)$, which is related to the force by
$F(x) = -dU(x)/dx$. The force, and hence the potential energy, usually
depend only on the position x of the particle, and not on its velocity dx/dt.
One important exception is the electromagnetic force, which does depend
on the velocity. For the harmonic oscillator, a simple integration of $F(x)$
yields $U(x) = kx^2/2$.

One also introduces the kinetic energy $T(dx/dt) = m \left(dx/dt \right)^2 /2$, which
depends on the velocity dx/dt. It is readily shown that the sum $T(dx/dt) +
U(x)$, denoted by the energy E, remains constant as the particle moves.
This result is the important principle of the conservation of mechanical
energy for systems that are not subject to friction.

Another important generalization is the introduction of generalized coordinates and generalized velocities, denoted by q and dq/dt, respectively.
For many systems, the familiar Cartesian coordinates, x, y, z, are not the
most convenient. For example, polar coordinates, r and θ, are often more
convenient than x and y for studying motion in the plane. Note that the
coordinate θ does not have the dimensions of length, a feature common
among generalized coordinates. Similarly, the generalized velocity dq/dt
need not have the dimensions of cm/sec. From now on, we shall consider U
and T to be functions of q and dq/dt. For an n-dimensional system, there
are, of course, n coordinates q and n velocities dq/dt.

1.1.2   *Principle of least action*

Newton's law of mechanics is expressed above in differential form. However,
Newton's law can also be expressed in integral form. It is proved in books on
mechanics that if (q_1, t_1) and (q_2, t_2) are the two endpoints of the trajectory
of a particle, then the complete trajectory will be that curve for which the
action $S(q)$ is a minimum, where $S(q)$ is given by the following integral
over time,

$$S(q) = \int_{t_1}^{t_2} L\left(q, \frac{dq}{dt} \right) \, dt, \tag{1.3}$$

where the Lagrangian $L(q, dq/dt)$ is defined by $L(q, dq/dt) = T(dq/dt) -
U(q)$.

The action $S(q)$ is a minimum if dS vanishes for an arbitrary infinitesimal change dq in the trajectory, keeping fixed the endpoints (q_1, t_1) and (q_2, t_2). Thus,

$$dS = 0 = S(q + dq) - S(q) = \int_{t_1}^{t_2} \left(\frac{\partial L}{\partial q} dq + \frac{\partial L}{\partial (dq/dt)} \frac{dq}{dt} \right) dt \qquad (1.4)$$

Integrating the second term by parts, and recalling that dL vanishes at the endpoints, leads to

$$\frac{\partial L}{\partial q} - \frac{d}{dt} \frac{\partial L}{\partial (dq/dt)} = 0, \qquad (1.5)$$

Equation (1.5) is the Lagrange's equation for each generalized coordinate q and generalized velocity dq/dt. Let us now apply the Lagrange's equation to the harmonic oscillator. The Lagrangian is $L(x, dx/dt) = T dx/dt - U(x) = \frac{m}{2}(dx/dt)^2 - \frac{1}{2}kx^2$. Therefore, the Lagrange's equation (1.5) indeed agrees with Newton's equation (1.1).

1.1.3 *Hamilton's equations*

Instead of the second-order differential equation (1.5) for the trajectory of the particle, one can transform the Lagrange's equation into two first-order differential equations, known as Hamilton's equations. This transformation not only decreases the difficulty of finding the trajectory of the particle, but Hamilton's equations are also a convenient point of departure for quantum mechanics.

The Lagrangian $L(q, dq/dt)$ is written in terms of the generalized coordinate q and generalized velocity dq/dt, whereas the Hamiltonian $H(q, p)$ is written in terms of the generalized coordinate q and generalized momentum p. The change in basis from $(q, dq/dt)$ to (q, p) is accomplished by a procedure known as a Legendre transformation. One introduces the Hamiltonian $H(q, p) = p(dq/dt) - L(q, dq/dt)$, where the generalized momentum p is given by $p = \partial L/\partial(dq/dt)$. Performing the canonical transformation from $L(q, dq/dt)$ to $H(p, q)$,

$$dL = \sum_i \left[\frac{\partial L}{\partial q_i} dq_i + \frac{\partial L}{\partial (dq_i/dt)} d(dq_i/dt) \right] =$$
$$\sum_i \left[\frac{dp_i}{dt} dq_i + p_i d(dq_i/dt) \right] = \qquad (1.6)$$
$$\sum_i \left[d(p_i dq_i/dt) + \frac{dp_i}{dt} dq_i - \frac{dq_i}{dt} dp_i \right]$$

one obtains

$$dH = \sum_i \left(\frac{dq_i}{dt} dp_i - \frac{dp_i}{dt} dq_i \right), \tag{1.7}$$

which can be rewritten as

$$\frac{\partial H}{\partial p_i} = \frac{dq_i}{dt}; \qquad \frac{\partial H}{\partial q_i} = -\frac{dp_i}{dt} \tag{1.8}$$

Let us check our results by comparing with the harmonic oscillator,

$$p = \partial L / \partial \left(dx/dt \right) = \partial T / \partial \left(dx/dt \right) = m \left(dx/dt \right) \tag{1.9}$$

which indeed corresponds to the momentum. It should be added that if the potential energy depends on the velocity, as is the case for the electromagnetic potential, then the generalized momentum p is not equal to the mechanical momentum $m \left(dx/dt \right)$ but contains an additional term, as we shall see.

Strictly speaking, one should take into account the possibility of an explicit time dependence in the Lagrangian $L(q, dq/dt, t)$ and in the Hamiltonian $H(q, p, t)$ to include the case in which the generalized coordinates are moving and thus depend explicitly on the time. And, of course, it should be understood that there are $3n$ generalized coordinates q_i and $3n$ generalized momenta p_i for a n-particle system.

Using (1.8), one can find the time derivative from an arbitrary function $f \left(q_1...q_n, p_1...p_n, t \right)$

$$\frac{d}{dt} f \left(q_i, p_i, t \right) = \frac{\partial f}{\partial t} + \sum_{i=1}^n \left(\frac{\partial f}{\partial q_i} \frac{dq_i}{dt} + \frac{\partial f}{\partial p_i} \frac{dp_i}{dt} \right) \equiv \frac{\partial f}{\partial t} + \{f, H\} \tag{1.10}$$

where $\{f, H\}$ is called the Poisson bracket. For two arbitrary functions $f \left(q_1...q_n, p_1...p_n, t \right)$ and $g \left(q_1...q_n, p_1...p_n, t \right),$ the Poisson bracket is

$$\{f, g\} = \sum_{i=1}^n \left(\frac{\partial f}{\partial q_i} \frac{\partial g}{\partial p_i} - \frac{\partial g}{\partial q_i} \frac{\partial f}{\partial p_i} \right) \tag{1.11}$$

Replacing $g(q_i, p_i, t)$ by $H(q_i, p_i, t)$ in (1.11) and evaluating the derivatives of $H(q_i, p_i, t)$ from Hamilton's equations (1.8) confirms equation (1.10).

1.1.4 *Hamilton-Jacobi equation*

In addition to the ordinary differential equations considered above, one can formulate the equation of motion in the form of a partial differential equation for the action function $S(q, t)$, where S is taken as a function of the upper limit of the integral in (1.3). The differential of Eq. (1.3) is

$$dS = \int_{t_1}^{t_2} \left[\frac{\partial L}{\partial q} - \frac{d}{dt} \frac{\partial L}{\partial (dq/dt)} \right] dqdt + \frac{\partial L}{\partial (dq/dt)} dq \bigg|_{t_1}^{t_2} = \qquad (1.12)$$

$$\frac{\partial L}{\partial (dq/dt)} dq \bigg|_{t_1}^{t_2}$$

where the Lagrange's equation (1.5) has been used. It follows from (1.3) and (1.12) that

$$\frac{dS}{dt} = L; \qquad \frac{\partial S}{\partial q} = \frac{\partial L}{\partial (dq/dt)} = p, \qquad (1.13)$$

which leads to the Hamilton-Jacobi equation,

$$\frac{\partial S}{\partial t} = L - p \frac{dq}{dt} = -H(q, p, t) \equiv -H \left(q, \frac{\partial S}{\partial q}, t \right) \qquad (1.14)$$

For the harmonic oscillator, the Hamilton function $H = p^2/2m + kx^2/2$, and, therefore, the Hamilton-Jacobi equation (1.14) takes the form

$$\frac{\partial S}{\partial t} + \frac{1}{2m} \left(\frac{\partial S}{\partial x} \right)^2 + \frac{1}{2} kx^2 = 0 \qquad (1.15)$$

When the Hamiltonian has no explicit time dependence, Eq. (1.15) reduces to an ordinary differential equation through the introduction of Hamiltonian characteristic function R, defined by $R = S + Et$,

$$\frac{1}{2m} \left(\frac{\partial R}{\partial x} \right)^2 + \frac{kx^2}{2} = E \qquad (1.16)$$

Equation (1.16) can be easily integrated to give R and S. Finally, the transformation equation $\partial S/\partial E = \alpha$ leads to

$$x = \sqrt{\frac{2E}{k}} \sin \left[\sqrt{\frac{k}{m}} (t + \alpha) \right] \qquad (1.17)$$

where the constant E is determined by the initial condition $E = p_0^2/2m + kx_0^2/2$. Equation (1.17) agrees with (1.2), as required.

1.2 Electrodynamics

1.2.1 *Maxwell's equations*

The time and space dependence of the electric and magnetic fields, $\mathbf{E}$ and $\mathbf{B}$, are given by Maxwell's equations, which are (in Gaussian units),

$$\nabla \times \mathbf{E} = -\frac{1}{c} \frac{\partial \mathbf{B}}{\partial t}; \qquad \nabla \mathbf{E} = 4\pi \rho \qquad (1.18)$$

$$\nabla \times \mathbf{B} = \frac{4\pi}{c}\mathbf{J} + \frac{1}{c}\frac{\partial \mathbf{E}}{\partial t}; \qquad \nabla \mathbf{B} = 0 \qquad (1.19)$$

where the sources of the fields are the electric charge density ρ and the electric current density $\mathbf{J}$. There are no magnetic charges or magnetic currents. There is a notational dispute regarding whether to call the magnetic field $\mathbf{B}$ or $\mathbf{H}$, and we will use the more common notation of $\mathbf{B}$.

The conservation of electric charge is contained in Maxwell's equations. Take the divergence of the first equation of (1.19) and the time derivative of the second equation of (1.18), recalling that the divergence of a curl vanishes. Combining these two results yields $d\rho/dt = -\nabla \mathbf{J}$. Integrating over volume V and replacing the volume integral of the divergence by a surface integral of the normal component of the current gives the desired result,

$$\frac{\partial}{\partial t}\int \rho \, dv = -\oint J_n \, dS \qquad (1.20)$$

Equation (1.20) states that the increase in charge in any volume of space equals the current that has entered (negative sign) through the surface enclosing that volume. Thus, electric charge is conserved.

Conservation of energy is also contained in Maxwell's equations. The fields produce energy density W and energy flux $\mathbf{\Phi}$, given by

$$W = \frac{E^2 + B^2}{8\pi}; \qquad \mathbf{\Phi} = \frac{c}{4\pi}\left(\mathbf{E} \times \mathbf{B}\right) \qquad (1.21)$$

Take the divergence of the cross product $\mathbf{E} \times \mathbf{B}$ which yields $\mathbf{B}\left(\nabla \times \mathbf{E}\right) - \mathbf{E}\left(\nabla \times \mathbf{B}\right)$, evaluate the curl from Maxwell's equation, and integrate over volume V. This yields the conservation of energy,

$$\frac{\partial}{\partial t}\int W \, dv = -\oint \Phi_n \, dS - \int \mathbf{J}\mathbf{E} \, dv \qquad (1.22)$$

Equation (1.22) states that the increase in energy in volume V equals the energy flux that has entered (negative sign) through the surface enclosing that volume minus the Joule heat $\mathbf{J}\mathbf{E}$. Thus, energy is conserved.

The energy flux $\mathbf{\Phi}$ is called the Poynting vector $\mathbf{P}$,

$$\mathbf{P} = \frac{c}{4\pi}\left(\mathbf{E} \times \mathbf{B}\right) \qquad (1.23)$$

Finally, the force exerted on charge e due to these fields is

$$\mathbf{F} = e\left[\mathbf{E} + \frac{1}{c}\left(\mathbf{v} \times \mathbf{B}\right)\right] \qquad (1.24)$$

where $\mathbf{v}$ is the velocity of the charged particle. Thus, the electromagnetic force depends on the velocity.

1.2.2 *Electromagnetic potentials*

In order to determine the Lagrangian that corresponds to the electromagnetic force, one must first determine the potential energy. Since the curl of the field $\mathbf{E}$ does not vanish, one cannot write $\mathbf{E}$ as the gradient of a scalar potential. However, the divergence of the field $\mathbf{B}$ does vanish. Therefore, one can write $\mathbf{B}$ as the curl of a vector $\mathbf{A}$, called the vector potential

$$\mathbf{B} = \nabla \times \mathbf{A} \tag{1.25}$$

Inserting (1.25) into the first of equations (1.18) yields

$$\nabla \times \left[\mathbf{E} + \frac{1}{c}\frac{\partial \mathbf{A}}{\partial t} \right] = 0 \tag{1.26}$$

Since the curl of the quantity in parenthesis does vanish, this quantity can be written as the (negative) gradient of a scalar function ϕ, called the scalar potential. In terms of the scalar and vector potentials,

$$\mathbf{E} = -\nabla\phi - \frac{1}{c}\frac{\partial \mathbf{A}}{\partial t} \tag{1.27}$$

Now that we have explicit expressions for $\mathbf{E}$ and $\mathbf{B}$ in terms of $\mathbf{A}$ and ϕ, we can insert (1.25) and (1.27) into Maxwell's equations to find the equations that are satisfied by $\mathbf{A}$ and ϕ,

$$\nabla^2 \mathbf{A} - \frac{1}{c^2}\frac{\partial^2 \mathbf{A}}{\partial t^2} = -\frac{4\pi}{c}\mathbf{J}; \quad \nabla^2\phi - \frac{1}{c^2}\frac{\partial^2 \phi}{\partial t^2} = -4\pi\rho \tag{1.28}$$

In the absence of both charges and currents, both $\mathbf{A}$ and ϕ satisfy the wave equation, as do both $\mathbf{B}$ and $\mathbf{E}$, whose solution is

$$\mathbf{A} = \mathbf{A}_0 \exp\left[i\left(\omega t - \mathbf{kr}\right)\right] + c.c. \tag{1.29}$$

leading (for $\nabla\phi = 0$) to

$$\mathbf{E} = -\frac{i\omega}{c}\mathbf{A}_0 \exp\left[i\left(\omega t - \mathbf{kr}\right)\right] + c.c. \tag{1.30}$$
$$\mathbf{B} = -\left(\mathbf{k} \times \mathbf{A}_0\right)\exp\left[i\left(\omega t - \mathbf{kr}\right)\right] + c.c$$

The absolute value of the Poynting vector, averaged over an oscillation of the field is then

$$\langle \mathbf{P} \rangle = \frac{\omega^2}{2\pi c}\left|\mathbf{A}_0\right|^2 \tag{1.31}$$

It should be mentioned that the above definitions of $\mathbf{A}$ and ϕ are not unique. From (1.25) and (1.27), it follows that Maxwell's equations are unchanged if one adds to $\mathbf{A}$ the gradient of any scalar function, say $\chi\left(\mathbf{r}, t\right)$, and also adds to ϕ the term $(1/c)\left(\partial\chi/\partial t\right)$. This feature of the potentials is called gauge invariance. The freedom to choose the function $\chi(\mathbf{r}, t)$ allows one to choose the set $(\mathbf{A}, \phi)$ such that

$$\nabla \mathbf{A} + \frac{1}{c}\frac{\partial \phi}{\partial t} = 0 \tag{1.32}$$

A choice of the function $\chi(\mathbf{r}, t)$ satisfying (1.32) is called a Lorentz gauge. A particularly useful choice of gauge in the absence of charges and currents is the Coulomb gauge, in which

$$\nabla \mathbf{A} = 0 \text{ and } \phi = 0 \tag{1.33}$$

The electromagnetic force can be written in terms of the vector and scalar potentials,

$$\mathbf{F} = e\left[-\nabla\phi - \frac{1}{c}\frac{\partial \mathbf{A}}{\partial t} + \frac{1}{c}\left(\mathbf{v} \times \nabla \times \mathbf{A}\right)\right] \tag{1.34}$$

One can rewrite this expression in a more convenient form,

$$F_x = -\frac{\partial U}{\partial x} + \frac{d}{dt}\left(\frac{\partial U}{\partial v_x}\right) \tag{1.35}$$

by introducing the velocity-dependent potential $U(\mathbf{r}, \mathbf{v})$ for the electromagnetic force,

$$U = e\phi - \frac{e}{c}\left(\mathbf{v}\mathbf{A}\right) \tag{1.36}$$

The corresponding Lagrangian is given in the usual way by

$$L = T - U = T - e\phi + \frac{e}{c}\left(\mathbf{v}\mathbf{A}\right) \tag{1.37}$$

The generalized momentum is the derivative of the Lagrangian with respect to velocity. The first term in (1.37) gives the usual mechanical momentum. The second term does not depend on velocity; the third term is new and gives $e\mathbf{A}/c$. Thus, the total generalized momentum is

$$\mathbf{p} = m\mathbf{v} + \frac{e\mathbf{A}}{c} \tag{1.38}$$

However, the Hamiltonian H, which gives the total energy, depends only on the mechanical momentum $m\mathbf{v}$. This follows from the definition of H,

$$H = \mathbf{p}\mathbf{v} - L \tag{1.39}$$

where $\mathbf{p}$ in (1.39) is the generalized momentum. One sees from (1.38) that the first term in (1.39), namely $\mathbf{p}\mathbf{v}$, contains an extra contribution. However, it follows from (1.37) that the second term of (1.39), namely $-L$, also contains an extra contribution. These two contributions exactly cancel and thus H remains unchanged,

$$H = T + e\phi \tag{1.40}$$

where the kinetic energy $T = mv^2/2$, as usual. Thus, the Hamiltonian (1.40), and hence the total energy, does not contain the magnetic field. This result is closely related to the fact that the magnetic field does not increase the velocity of the particle, but only changes its direction. The magnetic field causes the particle to rotate around the flux lines without any change in velocity, and hence no change in kinetic energy.

In terms of the generalized momentum, (1.38), the Hamiltonian $H\left(\mathbf{q},\mathbf{p}\right)$ has the form

$$H = \frac{1}{2m} \left(\mathbf{p} - \frac{e}{c}\mathbf{A}\right)^2 + e\phi \tag{1.41}$$

1.2.3 *Lagrangian for the electromagnetic field*

The Lagrangian discussed in the previous section was the function whose Lagrange's equations describe the motion of a charged particle subject to an electromagnetic force. However, one can also construct a Lagrangian density whose Lagrange's equations yield Maxwell's equations for the fields. For this Lagrangian density, the generalized coordinates are the scalar potential ϕ and the vector potential $\mathbf{A}$. The time derivatives of these generalized coordinates give the generalized velocities. However, there is a new feature here. The generalized coordinates ϕ and $\mathbf{A}$ also have a spatial dependence, $\partial\phi/\partial x$ and $\partial A_x/\partial x$ for each component of $\mathbf{A}$, and similarly for y and z. This spatial dependence leads to a new term in the Lagrange's equations, which now become

$$\frac{\partial L}{\partial q} - \frac{d}{dt}\frac{\partial L}{\partial\left(dq/dt\right)} - \nabla\frac{\partial L}{\partial\left(\nabla q\right)} = 0 \tag{1.42}$$

The appropriate Lagrangian density is

$$L = \frac{E^2 - B^2}{8\pi} - \rho\phi + \frac{1}{c}\mathbf{J}\mathbf{A} \tag{1.43}$$

in which $\mathbf{E}$ and $\mathbf{B}$ must be expressed in terms of $\mathbf{A}$ and ϕ according to (1.25) and (1.27).

We illustrate the procedure for the generalized coordinate ϕ. This coordinate appears only in the second term, whose derivative gives $-\rho$. The time dependence $d\phi/dt$ does not appear at all in the Lagrangian, and thus the second term in Eq. (1.42) vanishes. However, the spatial derivative

$\partial\phi/\partial x$ does appear in the third term through $E^2 = \left[\nabla\phi + (1/c)\left(\partial\mathbf{A}/\partial t\right)\right]^2$. Differentiating E^2 with respect to $\partial\phi/\partial x$ yields $-2E_x$, with similar contributions $-2E_y$ and $-2E_z$ arising from differentiating E^2 with respect to $\partial\phi/\partial y$ and $\partial\phi/\partial z$, respectively. Inserting these results into the Lagrange's equation yields $1/4\pi\sum_i\left(\partial E_i/\partial x_i\right) - \rho = 0$, which is the Maxwell equation $\nabla\mathbf{E} = 4\pi\rho$. The other Maxwell equations can be obtained in a similar manner.

One can obtain the Hamiltonian from the Lagrangian. One begins with the definition of the generalized momentum,

$$p_x = \frac{\partial L}{\partial\left(\partial A_x/\partial t\right)} = \frac{1}{4\pi}\left(\frac{1}{c}\frac{\partial A_x}{\partial t} + \frac{\partial\phi}{\partial x}\right) \qquad (1.44)$$

and similarly for p_y and p_z. The Hamiltonian $H\left(q,p\right) = pdq/dt - L$ then gives

$$H = 2\pi c^2 p^2 + \frac{1}{8\pi}\left(\nabla\times\mathbf{A}\right)^2 - cp\nabla\phi + \rho\phi - \frac{1}{c}\mathbf{J}\mathbf{A} \qquad (1.45)$$

Maxwell's equations follow from (1.45) upon applying Hamilton's equations (1.8).

1.3 Quantum mechanics

1.3.1 *Wave function*

One can establish a link between classical and quantum mechanics by comparing the general approach for studying macroscopic and microscopic objects. In classical mechanics, the motion of a particle of mass m, $\mathbf{r} = \mathbf{r}\left(t\right)$, subject to an external force $\mathbf{F}$, is defined by Newton's law, $md^2\mathbf{r}/dt^2 = \mathbf{F}$. The solution of this second-order differential equation is completely determined by the initial values of the position, $\mathbf{r}\left(t = 0\right) \equiv \mathbf{r}_0$, and velocity, $d\mathbf{r}/dt\left(t = 0\right) \equiv \left(d\mathbf{r}/dt\right)_0$ of the particle. Such a description cannot be used in quantum mechanics because of the central quantum mechanical postulate, the uncertainly principle $\Delta\mathbf{r}\Delta\mathbf{p} \sim h$. This principle states that exact knowledge of the position of a particle, $\Delta\mathbf{r} = 0$, implies that $\Delta\mathbf{p} = \infty$, i.e., the momentum $\mathbf{p}$ (or the velocity $d\mathbf{r}/dt$) remains completely unknown. Instead of the position and momentum, the state of a system in quantum mechanics is given by the wave function $\Psi\left(\mathbf{r},t\right)$, where $\left|\Psi^2\left(\mathbf{r},t\right)\right| d\mathbf{r}$ determines the probability for the particle to be found in the interval $\left(\mathbf{r},\mathbf{r} + d\mathbf{r}\right)$ at time t.

In quantum mechanics, to each physical quantity corresponds an appropriate operator: $i\hbar\,\partial/\partial t$ for the energy E, $-i\hbar\nabla$ for the momentum, etc. Accordingly, the classical law for a free particle, $E = p^2/2m$, is replaced by the quantum equation of motion (Schrödinger equation),

$$i\hbar\frac{\partial\Psi}{\partial t} = -\frac{\hbar^2}{2m}\nabla^2\Psi \qquad (1.46)$$

For a state described by the Hamiltonian H, one gets

$$i\hbar\frac{\partial\Psi}{\partial t} = H\Psi; \quad \Psi(t) = \Psi(0)\exp\left(-\frac{iHt}{\hbar}\right) \qquad (1.47)$$

If the function $\Psi(\mathbf{r})$ is one of the eigenfunctions $\varphi_n(\mathbf{r})$ of the operator $\hat{A}$,

$$\hat{A}\varphi_n = a_n\varphi_n, \qquad (1.48)$$

then in the state Ψ, the physical quantity A will definitely have the value a_n. Otherwise, there is the probability $|c_n|^2$ that in the state Ψ, a measurement of the quantity A will yield the value a_n, where c_n is given by expanding the function $\Psi(\mathbf{r})$ in a series of the eigenfunctions φ_n, $\Psi(\mathbf{r}) = \sum_n c_n\varphi_n(\mathbf{r})$.

1.3.2 *Dynamic behavior*

For the normalized function Ψ, $\int \Psi^*(\mathbf{r},t)\Psi(\mathbf{r},t)\,d\mathbf{r} = 1$, one obtains from Eq. (1.48) with ϕ_n denoted by Ψ,

$$a_n = \int \Psi^*(\mathbf{r},t)\hat{A}\Psi(\mathbf{r},t)\,d\mathbf{r} \qquad (1.49)$$

Differentiating this equation and using Eq. (1.47) yields

$$\frac{da_n}{dt} = \int \Psi^*(\mathbf{r},t)\left[\frac{\partial\hat{A}}{\partial t} + \frac{i}{\hbar}\left(\hat{A}\hat{H} - \hat{H}\hat{A}\right)\right]\Psi(\mathbf{r},t)\,d\mathbf{r} \qquad (1.50)$$

Therefore, the square brackets in (1.50) represent the operator for the total time derivative da_n/dt. This operator contains the quantum Poisson brackets, which bear a close resemblance to the classical Poisson brackets defined in Eq. (1.10). A comparison between classical and quantum Poisson brackets shows that they stand in the following one-to-one correspondence,

$$\{f,g\}_{class} \leftrightarrow -\frac{i}{\hbar}(f,g)_{quantum} \qquad (1.51)$$

One can confirm this correlation if f and g are the coordinate and momentum, respectively. The classical Poisson brackets (1.10) immediately yields unity. In the quantum Poisson brackets (1.51), replacing g by the quantum mechanical operator $-i\hbar\partial/\partial x$ one also yields unity.

1.3.3　*Conservation laws in quantum mechanics*

In classical mechanics, the conservation laws of energy, momentum and angular momentum follow from the homogeneity of time and the homogeneity and isotropy of space. This raises the question of the validity of these conservation laws in quantum mechanics. Moreover, in quantum mechanics there are additional conservation laws connected with parity and the indistinguishability of identical particles.

If there are no external fields, and the Hamiltonian does not depend explicitly on time, $\partial H / \partial t = 0$, the conservation of energy follows directly from Eq. (1.50). As in classical mechanics, the conservation of momentum follows from the homogeneity of space. For a system not subject to an external field, all points in space are equivalent. Therefore, a shift of all coordinates from $\mathbf{r}_i$ to $\mathbf{r}_i + \delta \mathbf{r}$ does not change the Hamiltonian of the system. However, such a displacement transforms the wave function $\Psi(\mathbf{r}_1, \mathbf{r}_2 ...)$ into

$$\Psi(\mathbf{r}_1 + \delta \mathbf{r}, \mathbf{r}_2 + \delta \mathbf{r} ...) = \Psi(\mathbf{r}_1, \mathbf{r}_2 ...) + \delta \mathbf{r} \sum_i \nabla_i \Psi \qquad (1.52)$$

$$= \left(1 + \delta \mathbf{r} \sum_i \nabla_i \right) \Psi(\mathbf{r}_1, \mathbf{r}_2 ...)$$

Due to the homogeneity of space, the expression $\left(1 + \delta \mathbf{r} \sum_i \nabla_i \right)$ and, therefore, the operator ∇_i commutes with the Hamiltonian H. To obtain consistency with Eq. (1.51), one defines the conserved momentum operator as $-i\hbar\nabla$. Similarly, the equivalence of all directions in space (isotropy of space), implies that the Hamiltonian cannot change when the entire system is rotated through an angle $\delta\phi$, so that $\delta \mathbf{r}_i = \delta\phi \times \mathbf{r}_i$. Hence, analogous to (1.52), the operator corresponding to such a rotation is

$$1 + \sum_i (\delta\phi \times \mathbf{r}_i) \nabla_i = 1 + \delta\phi \sum_i (\mathbf{r}_i \times \nabla_i) \qquad (1.53)$$

The quantity whose conservation follows from the isotropy of space is the angular momentum, and its operator $\hat{L}$ is thus

$$\hat{L} = -i\hbar \mathbf{r} \times \nabla = \mathbf{r} \times \hat{\mathbf{p}} \qquad (1.54)$$

As in the case of linear momentum, the factor $-i\hbar$ is chosen for consistency with (1.51).

Thus far, a conservation laws in quantum mechanics are consistent with analogous conservation laws in classical mechanics. The first specifically quantum mechanical conservation law, that does not has a classical counterpart is the conservation of parity. In addition to translation and rotation in space, there is another operation, $\hat{P}$, that leaves invariant the Hamiltonian operator of a closed system. This operation is inversion, i.e., the simultaneous change of the sign of all spatial coordinates, $\mathbf{r} \rightarrow -\mathbf{r}$,

$$\hat{P}\Psi(\mathbf{r}) = \Psi(-\mathbf{r}) \tag{1.55}$$

Applying $\hat{P}$ twice to (1.55) leaves the wave function unchanged, $\hat{P}^2\Psi(\mathbf{r}) = \Psi(\mathbf{r})$. Therefore, the effect of $\hat{P}$ is to multiply $\Psi(\mathbf{r})$ by $+1$ or by -1. All particles are thus divided into two groups, those for which the wave function does not change upon inversion and those for which the wave function does change sign upon inversion. These two groups of particles are said to have even or odd parity, respectively. Since the operator $\hat{P}$ commutes with the Hamiltonian $\hat{H}$, this property is conserved in time. The conservation of parity is important in the matrix formulation of quantum mechanics, where instead of operators $\hat{A}$, one uses the matrix elements A_{mn}, where A_{mn} correspond to the transition from state n to state m under the influence of operator $\hat{A}$,

$$A_{mn} = \int \Psi_m^*(\mathbf{r}) \hat{A}\Psi_n(\mathbf{r})\, d\mathbf{r} \tag{1.56}$$

If operator $\hat{A}$ has even parity, the transition probability A_{mn} can differ from zero only if states m and n have the same parity. If these states have different parity, the integrand in Eq. (1.56) will change sign under the inverse operation, and the integral will equal zero.

Another distinction between classical and quantum mechanics manifests itself in time-reversal symmetry. Newton's equations of motion involve a second-order time derivative, whereas the Schrödinger equation contains a first-order time derivative. Since the Schrödinger equation changes sign under the time-inversion operation $\hat{T}$, $t \rightarrow -t$, simultaneously with changing $\Psi(t)$ to $\Psi(-t)$, one has to change Ψ to Ψ^*,

$$\hat{T}\Psi(r,t) = \Psi^*(r,-t) \tag{1.57}$$

The operation of complex conjugation does not affect the position of the particles, since the particle density at any point is proportional to $\Psi\Psi^*$.

The concept of indistinguishability of particles, like their parity, is specific to quantum mechanics and has no classical analogue. In classical mechanics, one can always distinguish between different particles by specifying their different positions and velocities, which is not possible in quantum mechanics because of the uncertainty principle. In quantum mechanics, all particles of the same type are indistinguishable. Therefore, analogously to (1.55), interchanging the coordinates of each pair of particles, returns one to the original system. Because of this, there are two groups of particles (bosons and fermions), which do or do not change the sign of their wave function upon interchanging the coordinates of a pair of particles.

1.3.4 *Different representations*

Thus for, we have considered the Schrödinger representation, in which the eigenfunctions are functions of the time while the operators (in the absence of an explicit time dependence) do not depend on time. However, there is another representation, called the Heisenberg representation, in which the operators depend on time and the eigenfunctions are time-independent. This representation is obtained by inserting (1.47) into (1.49),

$$a_n = \int \Psi^* (0) \exp \left(\frac{iHt}{\hbar} \right) \hat{A} \exp \left(-\frac{iHt}{\hbar} \right) \Psi (0) \, d\mathbf{r} \qquad (1.58)$$

with the Heisenberg operator $\hat{A}_H$ related to the Schrödinger operator $\hat{A}_S$ by

$$\hat{A}_H = \exp \left(\frac{iHt}{\hbar} \right) \hat{A}_S \exp \left(-\frac{iHt}{\hbar} \right) \qquad (1.59)$$

Finally, for the Hamiltonian $H = H_0 + H_I$, with interaction energy H_I, one can use the interaction representation, where both the operator $\hat{A}_I$ and the eigenfunctions Ψ_I depend on the unperturbed Hamiltonian H_0,

$$\hat{A}_I = \exp \left(\frac{iH_0t}{\hbar} \right) \hat{A}_S \exp \left(-\frac{iH_0t}{\hbar} \right) ; \ \Psi_I = \exp \left(\frac{iH_0t}{\hbar} \right) \Psi_S \qquad (1.60)$$

1.3.5 *Aharonov-Bohm effect*

Quantum mechanics gave birth to an interesting change in classical electrodynamics. The vector potential $\mathbf{A}$, which was introduced classically solely for mathematical convenience, takes on new meaning in quantum mechanics. The Aharonov–Bohm effect shows that one can affect measurements by passing an electron through a region of space that has zero $\mathbf{E}$ and zero $\mathbf{B}$ fields but non-zero $\mathbf{A}$. According to Section 1.2.2, the Schrödinger equation for a charged particle in an external magnetic field $\mathbf{B} = \nabla \times \mathbf{A}$ with $\phi = 0$, has the following form,

$$H\Psi = \left[\frac{1}{2m} \left(-i\hbar\nabla - \frac{e}{c}\mathbf{A} \right)^2 \right] \Psi = E\Psi \qquad (1.61)$$

If $\Psi_0(\mathbf{r})$ is the solution of the Schrödinger equation in the absence of a magnetic field, the solution of (1.61) is

$$\Psi(\mathbf{r}) = \Psi_0(\mathbf{r}) \exp\left(\frac{2ie}{\hbar c} \int^{\mathbf{r}} A(z)\ dz\right) \tag{1.62}$$

where the integration is performed over an arbitrary trajectory terminating at point $\mathbf{r}$. The validity of (1.62) can be easily verified by a double application of the operator $-i\hbar\nabla - e\mathbf{A}/c$ to Eq. (1.62).

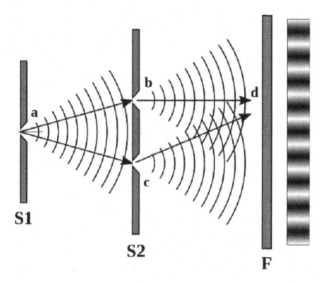

Fig. 1.1   Passing beam of particles through a screen with two slits (Stern-Gerlach experiment).

In Fig. 1.1, we display the well-known experimental set-up of a plane having two slits, which proves the particle-wave nature of quantum particles. A stream of particles passing through two slits exhibits on the screen the interference pattern typical of waves. Assume now that a long cylinder with an electric current $\mathbf{J}$ is placed behind the slits perpendicular to their plane. As the current and the corresponding magnetic field are changed, the interference pattern is shifted. If the cylinder is long enough, the magnetic field created by the current $\mathbf{J}$, remains inside the cylinder and cannot change the interference pattern. However, unlike the magnetic field, the vector potential $\mathbf{A}$ is nonzero everywhere outside the cylinder. This can be seen by considering the integral over a closed path surrounding the cylinder,

$$\oint \mathbf{A} dl = \int (\nabla \times \mathbf{A}) \, d\mathbf{S} = \int B_n dS \equiv \Phi, \tag{1.63}$$

where Φ is the magnetic flux. Therefore, integrating over the closed path in (1.62) shows that the wave function of the electron, which depends on the vector potential $\mathbf{A}$, is altered by the magnetic flux created by the current inside a cylinder.

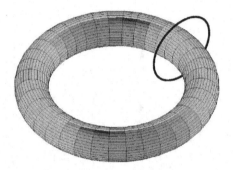

Fig. 1.2   Superconducting ring in a magnetic field.

An analogous effect occurs in a superconducting ring (Fig. 1.2), where the electron wave function has the form (1.62). Consider the integral over a closed contour around the ring starting at some point outside the ring. Since we return to the starting point, the wave function cannot change, i.e., the phase can change only by a multiple of 2π,

$$\frac{2e}{\hbar c} \oint A(z) \, dz \equiv \frac{4\pi e}{\hbar c} \Phi = 2\pi n \tag{1.64}$$

implying that the magnetic flux changes by an integral number of elementary units (fluxons), each equal to $hc/2e$.

Chapter 2

Semiclassical Theory of Radiation

In this Chapter we discuss the semiclassical theory of radiation. The theory is denoted "semiclassical" because the particle is treated quantum mechanically but we restrict ourselves to the classical description of the electromagnetic field. The full quantum analysis of both particle and field will be presented in Section 3.7. After presenting the general theory in Sections 2.1-2.4, we will consider three effects, each of which was awarded a Nobel prize (Einstein - 1905, Cherenkov - 1958, Compton - 1927).

2.1   Fermi's Golden Rule

According to Eq. (1.41), the Schrödinger equation of a particle interacting with an electromagnetic field described in the Coulomb gauge ($\nabla \mathbf{A} = 0, \phi = 0$) has the following form,

$$i\hbar \frac{\partial \Psi(\mathbf{r}, t)}{\partial t} = \frac{1}{2m} \left(-i\hbar\nabla - \frac{e}{c}\mathbf{A} \right)^2 \Psi \tag{2.1}$$

$$= \left(-\frac{\hbar^2}{2m}\nabla^2 + \frac{ie\hbar}{mc}\mathbf{A}\nabla + \frac{e^2}{2mc^2}A^2 \right) \Psi(\mathbf{r}, t)$$

One can solve Eq. (2.1) by means of perturbation theory, considering the term linear in $\mathbf{A}$ as a perturbation and neglecting the term quadratic in $\mathbf{A}$,

$$i\hbar \frac{\partial \Psi}{\partial t} = \left(-\frac{\hbar^2}{2m}\nabla^2\Psi + V \right)\Psi; \qquad V \equiv \frac{ie\hbar}{mc}\mathbf{A}\nabla \tag{2.2}$$

In the zero-order approximation, Eq. (2.1) reduces to (1.46) for a free particle having eigenvalues E_k^0 and the wave functions $\Psi_k(\mathbf{r}) = \Psi_{k,0}(\mathbf{r}) \exp\left(-itE_k^0/\hbar\right)$. One expands the wave function $\Psi(\mathbf{r}, t)$ in terms of the unperturbed eigenfunctions $\Psi_k(\mathbf{r})$,

$$\Psi(\mathbf{r},t) = \sum_k a_k(t)\,\Psi_{k,0}(\mathbf{r})\exp\left(-itE_k^0/\hbar\right) \tag{2.3}$$

Inserting (2.3) into (2.2) and using the properties of the unperturbed functions $\Psi_{k,0}(\mathbf{r})$ gives

$$i\hbar\frac{da_k}{dt} = \sum_l V_{kl}(t)\exp\left(i\omega_{lk}t\right)a_l(t) \tag{2.4}$$

where $\omega_{lk} = \left(E_l^0 - E_k^0\right)/\hbar$ and

$$V_{kl}(t) = \int \Psi_{k,0}^*(\mathbf{r},t)\,V\Psi_{l,0}(\mathbf{r},t)\,d\mathbf{r} \tag{2.5}$$

As is usually done in perturbation theory, one assumes that the system was initially in state k, so that $a_{k,0} = 1$ and $a_{l,0} = 0$ for all $l \neq k$. The perturbation starts to act at $t = 0$, and for $t > 0$, we seek a solution of Eq. (2.4) of the form

$$a_k = 1 + \varepsilon a_{k,1} + \varepsilon^2 a_{k,2} + \ldots \tag{2.6}$$
$$a_l = \varepsilon a_{l,1} + \varepsilon^2 a_{l,2} + \ldots \quad (l \neq k)$$

where the small parameter ε is proportional to the perturbation potential matrix element V_{kl}. To first order in perturbation theory, one obtains

$$a_{k,1} = -\frac{i}{\hbar}\int_0^t V_{kk}(\tau)\,d\tau; \quad a_{l,1} = -\frac{i}{\hbar}\int_0^t V_{kl}(\tau)\exp\left(i\omega_{lk}\tau\right)d\tau; \tag{2.7}$$

According to Eqs. (1.29) and (2.2), the matrix element of the perturbation is

$$V_{kl}(\tau) = F_{kl}\exp\left[i\left(\omega_{lk}-\omega\right)t\right] + F_{kl}^*\exp\left[i\left(\omega_{lk}+\omega\right)t\right] \tag{2.8}$$

where

$$F_{kl} = \frac{ie\hbar}{mc}\int \Psi_{k,0}^*(\mathbf{r})\exp\left(i\mathbf{kr}\right)\mathbf{A}_0\nabla\Psi_{l,0}(\mathbf{r})\,d\mathbf{r} \tag{2.9}$$

Inserting (2.8)-(2.9) into (2.7) leads to

$$a_{l,1} = -\frac{F_{kl}\left\{\exp\left[i\left(\omega_{lk}-\omega\right)t\right]-1\right\}}{\hbar\left(\omega_{lk}-\omega\right)} - \frac{F_{kl}^*\left\{\exp\left[i\left(\omega_{lk}+\omega\right)t\right]-1\right\}}{\hbar\left(\omega_{lk}+\omega\right)} \tag{2.10}$$

Considering only the first term in Eq. (2.10),

$$|a_{l,1}|^2 = \frac{4|F_{kl}|^2\sin^2\left[\left(\omega_{lk}-\omega\right)t/2\right]}{\hbar^2\left(\omega_{lk}-\omega\right)^2} \tag{2.11}$$

The second term in Eq. (2.10) leads to the same result with ω replaced by $-\omega$, which means that the probabilities are equal for the direct and reverse transitions between each pair of states.

The probability $|a_{l,1}|^2$ for a transition from state k to state l is appreciable only when one of the denominators is close to zero, which describes the absorption of the quantum of the electromagnetic field, $E_l^0 = E_k^0 + \hbar\omega$, or its emission, $E_l^0 = E_k^0 - \hbar\omega$. However, to get finite values for the probability of the transition, one has to take into account the width of energy levels or consider non-monochromatic radiation. Let the final state l be continuous with energy density of states $\rho(\varepsilon)$. Then, the probability of transition per the unit time W is

$$W = \frac{1}{t} \int \left| a_{l,1} \left(\frac{\varepsilon}{\hbar} \right) \right|^2 \rho(\varepsilon)\, d\varepsilon \qquad (2.12)$$

Since the integrand in Eq. (2.12) has a sharp maximum centered around $\omega_{lk} = \omega$, one can consider $|F_{kl}|^2$ as nearly constant and take this factor out of the integral. Noting that

$$\int_{-\infty}^{\infty} \frac{\sin^2 x}{x^2}\, dx = \pi, \qquad (2.13)$$

one obtains Fermi's Golden Rule,

$$W = \frac{2\pi}{\hbar} \rho(\varepsilon_l) |F_{kl}|^2 \delta(\omega_{lk} - \omega) \qquad (2.14)$$

Another possibility for dealing with the divergence in Eq. (2.10) is to replace the monochromatic radiation by radiation with intensity $I(\omega)\Delta\omega$ in the small frequency interval $\Delta\omega$. The intensity (averaged Poynting vector, $\langle P \rangle$) is defined in Eq. (1.31) through the square of $\mathbf{A}_0$ as

$$|\mathbf{A}_0|^2 = \frac{2\pi c}{\omega^2} I(\omega) \qquad (2.15)$$

Writing separately the factor $|\mathbf{A}_0|^2$ in Eq. (2.11) and performing calculations analogous to (2.11)-(2.14), one obtains

$$W = \frac{4\pi^2 e^2}{cm^2\omega_{lk}^2} I(\omega_{lk}) \left| \int \Psi_{k,0}^*(\mathbf{r}) \exp(i\mathbf{kr}) \nabla_A \Psi_{l,0}(\mathbf{r})\, d\mathbf{r} \right|^2 \qquad (2.16)$$

2.2 Dipole transitions

In many cases, the wavelength of the radiation is much larger than size of the system. For instance, in the scattering of visible light by atoms, the light wavelength is of order 10^{-4} cm while the size of atoms is of order 10^{-8} cm, i.e. the factor $kr = 2\pi r/\lambda$ in (2.9) is much smaller than unity. Therefore, one can replace $\exp(i\mathbf{kr})$ by unity,

$$F_{kl} = \frac{ie\hbar}{mc} \int \Psi^*_{k,0}(\mathbf{r})\, \mathbf{A}_0 \nabla \Psi_{l,0}(\mathbf{r})\, d\mathbf{r} = -\frac{eA_0}{mc}\,(p_A)_{kl} \qquad (2.17)$$

where $(p_A)_{kl}$ is the matrix element of the projection of the momentum in the direction of the polarization of the electromagnetic field. According to Eq. (1.50), the time dependence of the velocity $d\mathbf{r}/dt$ is defined as

$$\frac{d\hat{\mathbf{r}}}{dt} = \frac{i}{\hbar}\left(\hat{H}\hat{\mathbf{r}} - \hat{\mathbf{r}}\hat{H}\right) = \frac{i}{2m\hbar}\left(\hat{\mathbf{p}}\hat{\mathbf{p}}\hat{\mathbf{r}} - \hat{\mathbf{r}}\hat{\mathbf{p}}\hat{\mathbf{p}}\right) = \frac{i}{2m\hbar}\left(-2i\hbar\hat{\mathbf{p}}\right) = \frac{\hat{\mathbf{p}}}{m} \qquad (2.18)$$

and $(p_A)_{kl} = m\,(d\mathbf{r}/dt)_{kl} = im\omega\,(\mathbf{r})_{kl}$. Since such transitions are proportional to the matrix elements of the dipole moment, $e\mathbf{r}$, they are called dipole transitions.

2.3   Forbidden and strictly forbidden transitions

In the previous section we considered dipole transitions, which arise from the first term in the series expansion of $\exp(i\mathbf{kr}) \approx 1 + i\mathbf{kr} + \dots$. Because of the symmetries of the k and l states, the dipole moment $(r)_{kl}$ may exactly vanish and one has to take into account the next terms in the expansion, which leads to quadrupole and higher-order transitions. Even the first non-zero term, $i\mathbf{kr}$, results in a non-zero but very small $(\mathbf{kr})^2$ contribution to the transition probability. Such transitions between k and l states are called forbidden. Moreover, for some symmetries, the matrix element (2.9) vanishes. For example, if both states, $\Psi^*_{k,0}(\mathbf{r})$ and $\Psi_{l,0}(\mathbf{r})$, are spherically symmetric, one can choose the $x-$axis in the direction of $\mathbf{A}$, and rewrite Eq. (2.9) in the following form,

$$F_{kl} = \frac{ie\hbar}{mc} \int \Psi^*_{k,0}(\mathbf{r})\exp\left[i\,(k_y y + k_z z)\right] A_0 \frac{\partial}{\partial x}\Psi_{l,0}(\mathbf{r})\, d\mathbf{r} \qquad (2.19)$$

Since the integrand in Eq. (2.19) is odd in x, the integral vanishes. However, even in this case (strictly forbidden transitions), the transition probability is not identically zero, because we took into account in Eq. (2.1) only terms linear in $\mathbf{A}$. The quadratic terms in $\mathbf{A}$ will also make a very small contribution to the transition probability.

2.4 Selection rules

Selection rules give the pairs of states of a system between which transitions are allowed when the system interacts with the electromagnetic field. For example, the non-zero matrix element $(\mathbf{r})_{kl}$ allows dipole transitions. If the electromagnetic field propagates along the z-axis, $z = r\cos\theta = rY_{1,0}(\theta)$, one can write the matrix element $(\mathbf{r})_{kl}$ in the following form,

$$(\mathbf{r})_{kl} = \int \Psi_N^*(r) Y_{l,m}^*(\theta,\phi) \mathbf{r} Y_{1,0}(\theta) \Psi_{N'}(r) Y_{l',m'}(\theta,\phi)$$
$$\times r^2 dr \sin\theta d\theta d\phi, \qquad (2.20)$$

where the $Y_{l,m}(\theta,\phi)$ are the normalized spherical harmonics.

Using the well-known properties of the $Y_{l,m}(\theta,\phi)$ [10], one readily finds that the integral (2.20) is non-zero only if

$$l' = l \pm 1; \qquad m' = m \qquad (2.21)$$

The preceding results follow from the linear polarization of the electromagnetic field, $A = A_z$. In the case of spherical polarization, $A = A_x \pm iA_y$, the analogous analysis leads to the following selection rules

$$l' = l \pm 1; \qquad m' = m \pm 1 \qquad (2.22)$$

2.5 Radioactivity

Most of the elements in the periodic table have isotopes that are unstable. This spontaneous radioactive decay results in the formation a more stable atom. An example is the behavior of carbon C_{12}. The nuclei of all isotopes of carbon have the same number (six) of protons, but they differ in the numbers of neutrons. The nucleus of the isotope C_{14} has an excess of neutrons, and it decays by emitting an electron (negatively charged beta particle), whereas the isotope C_{11} which has a deficiency of neutrons, decays by emitting a positron (positively charged beta particle). The particles resulting from radioactive decay are alpha particles and beta particles. An alpha particle is a doubly-charged nucleus of helium, consisting of two protons and two neutrons. After alpha decay, the radioactive atom moves two places down on the periodic table. Alpha decay is relatively rare because it requires a lot of energy to form an alpha particle from two protons and two neutrons. Beta particles are negative or positive singly-charged particles (electrons or positrons). The third type of radioactive decay is the emission of very short wavelength gamma rays, which, being photons, have zero mass. Gamma rays are emitted together with alpha and beta particles, when the nucleus decays from an excited state to the ground state. In contrast to alpha and beta decay, the nucleus loses energy during gamma decay without changing its character.

The radioactive decay rate W is given by Fermi's Golden Rule (2.14). The mean lifetime τ of a system is defined as $\tau = 1/W$. Consider the theory of beta decay for the process of neutron-proton transformation,

$$n \to p + e^- + \nu \tag{2.23}$$

where ν is a neutrino, whose existence was first postulated by Pauli in 1930 to explain the energies of the electrons emitted in beta decay. As explained earlier, to get the value of W, one has to consider the transition from an initial state of definite energy to a final state with a range of final states having a energy density of states $\rho\left(E_i\right)$. Assuming that the final state of the proton has negligible kinetic energy, one can write the conservation laws for energy and momentum in the following form,

$$E_i = \varepsilon + cq; \qquad \mathbf{P} + \mathbf{p} + \mathbf{q} = 0 \tag{2.24}$$

where $\mathbf{P}$ is the momentum of the decayed nucleus, and ε, $\mathbf{p}$ and cq, $\mathbf{q}$ are the energy and momentum of the electron and neutrino, respectively.

The density of states for the electron-neutrino system is

$$\rho\left(E_i\right) = \frac{4\pi q^2 dq}{(2\pi\hbar)^3} \frac{4\pi p^2 dp}{(2\pi\hbar)^3} = \frac{\left(E_i - \varepsilon\right)^2}{c^3} \frac{p^2 dp}{4\pi^4\hbar^6} \tag{2.25}$$

where the conservation of energy has been used in the last equality.

A slightly modified version of the foregoing calculation will be used in the next Sections for the analysis of the interaction of a gamma ray with an electron (Compton effect), and a photon ejecting an electron from an atom (photoelectric effect).

2.6   Photoelectric effect

If the electric charge is rigidly bound in a system, such as an electron in an atom or a proton in a nucleus, the strong electromagnetic field is able to destroy the system by decoupling the bound charge into a free state (photoelectric effect for atoms and photodissociation for nuclei). This process occurs if the energy of the electromagnetic field $\hbar\omega$ is larger than the binding energy ε_0 of the charge, leading to

$$\frac{\hbar^2 k^2}{2m} = \hbar\omega - \varepsilon_0 \tag{2.26}$$

This important formula not only confirms the particle nature of the field, but also shows that the energy of a free charge $\hbar^2 k^2/2m$ depends on the frequency of the electromagnetic field and not on its intensity. The latter determines the number of charges knocked from the system, but not their energy.

As an example, we apply Fermi's Golden Rule (2.14) to the calculation of the probability of the transition of the $1s$ electron in an atom to the free state. The wave functions for the $1s$ electron and the free electron are,

$$\Psi_{1s} = \frac{1}{\sqrt{\pi a_B^3}} \exp\left(-\frac{r}{a_B}\right); \qquad \Psi = \exp\left(i\mathbf{q}\mathbf{r}\right) \tag{2.27}$$

where a_B is the Bohr radius, $a_B = \hbar^2/me^2$.

Inserting (2.27) into the matrix element (2.9) gives

$$F_{kl} = C \int \exp\left(-i\mathbf{q}\mathbf{r}\right) \exp\left(i\mathbf{k}\mathbf{r}\right) \mathbf{A}_0 \nabla \exp\left(-\frac{r}{a_B}\right) r^2 dr \sin\theta d\theta \tag{2.28}$$

where $C = (2ie\hbar/mc)\sqrt{\pi/a_B^3}$.

Integrating (2.28) by parts and performing the integrations over the angle θ and over r leads to

$$F_{kl} = C\left(\mathbf{A}_0\mathbf{q}\right) \int \frac{\sin\left(Kr\right)}{K} \exp\left(-\frac{r}{a_B}\right) r dr = C\frac{\left(\mathbf{A}_0\mathbf{q}\right)}{\left(1 + K^2 a_B^2\right)^2} \tag{2.29}$$

where $\hbar\mathbf{K} = \hbar\left(\mathbf{k} - \mathbf{q}\right)$ is the momentum transferred to the atom. The transition probability W for the photoelectric effect is

$$W \propto F_{kl}^2 \propto \frac{\left(\mathbf{A}_0\mathbf{q}\right)^2}{\left(1 + K^2 a_B^2\right)^4} \tag{2.30}$$

According to Eq. (2.30), the probability is greatest when the emitted electron propagates in the direction of the polarization $\mathbf{A}_0$ of the field, i.e., in the direction of the electric component of the field.

2.7 Cherenkov effect

The speed of light in vacuum, $c = 3 \times 10^{10}$ cm/sec, is the maximum possible velocity for material particles. However, in a medium the speed of light

$v = c/n$, where n is the refractive index. For example, the speed of light in water is about $0.7c$. Therefore, it is possible that the velocity of the light source in a medium will be greater than the velocity of the light in this medium. When this occurs for sound waves, the Mach effect means that one first sees the fast moving source of sound (lightning or an airplane), and only afterwards does one hear the sound. The similar effect for light is called the Cherenkov effect.

A moving charge emits energy in the form of spherical waves propagating in all directions. However, if the velocity of the charge is greater than the velocity of light in the medium, the wave "cloud" in each succeeding moment of time will be closer and closer to the charge. As this takes place, the boundary of radiation forms the Mach cone, which is inclined at an angle θ, where $\cos\theta = c/vn$. Combining this measurement with the motion of a charge in a magnetic field, one can find the mass of the particle. The energy transferred to the medium leads to the ionization of the medium, giving off blue radiation. This radiation was first measured by Cherenkov, and is known as the Cherenkov effect. This effect is responsible for the blue glow of the water that surrounds the core of a nuclear reactor.

2.8 Compton effect

The shift in the wavelength of light scattered from stationary electrons, discovered by Compton, was an important proof of the Planck hypothesis of photons. Consider a collision between a quantum of the electromagnetic field and a stationary electron. The electromagnetic field has the energy $\hbar\omega$ and momentum $\hbar k$ before a collision, and $\hbar\omega'$ and $\hbar k'$ after a collision, and a stationary electron has relativistic energy mc^2 before a collision and $\sqrt{p^2c^2 + m^2c^4}$ after a collision, where $\mathbf{p}$ is the electron momentum after the collision. Conservation of momentum and energy imply

$$\hbar\mathbf{k} - \hbar\mathbf{k}' = \mathbf{p} \tag{2.31}$$

$$\hbar\omega + mc^2 = \hbar\omega' + \sqrt{p^2c^2 + m^2c^4}$$

Comparing the two expressions for p^2 in Eqs. (2.31) yields

$$\left(\hbar\omega - \hbar\omega' + mc^2\right)^2 - mc^4 = \left(\hbar\omega\right)^2 + \left(\hbar\omega'\right)^2 - 2\left(\hbar\omega\right)\left(\hbar\omega'\right)\cos\theta \tag{2.32}$$

where θ is the scattering angle for the incident light. Simplifying (2.32) and introducing the wavelength $\lambda = c/\omega$, leads to

$$\lambda' - \lambda = \frac{\hbar}{mc}\left(1 - \cos\theta\right) \tag{2.33}$$

where $\hbar/mc = 0.025$ Å is called the Compton wavelength. Since the maximum shift in wavelength is only 0.05 Å, the Compton effect requires the scattering of x-ray photons.

2.9 Problems

Problem 2.1.

Find the selection rules for magnetic-dipole and electric-quadrupole transitions by an electron in a centrally symmetric potential.

Solution.

According to Fermi's Golden Rule, the probability for scattering is proportional to the square of the matrix element of the scattering potential between the initial and final electron states. In the present case, to lowest order in the fields, the scattering potential is proportional to $\mathbf{A}\nabla$. If one writes the electromagnetic field in the gauge in which the scalar potential vanishes and the vector potential satisfies $\nabla\mathbf{A} = 0$, we have seen that $\mathbf{A}(\mathbf{r}, t)$ satisfies the wave equation and hence $\mathbf{A}(r) \sim \mathbf{A}_0 \exp(i\mathbf{k}\mathbf{r})$, where the direction of $\mathbf{A}_0$ gives the polarization of the wave. Therefore, $\mathbf{A}\nabla$ is proportional to $\mathbf{A}_0 \exp(i\mathbf{k}\mathbf{r})\nabla_{\mathbf{A}}$, where the gradient is taken in the direction of the polarization of the wave, which will be chosen as the z-direction. Upon integrating by parts, the derivative with respect to z is replaced by $z = r\cos\theta$.

Expanding the exponential $\exp(i\mathbf{k}r)$ for small $\mathbf{k}$ gives

$$exp(i\mathbf{k}\mathbf{r}) \approx 1 + i\mathbf{k}\mathbf{r} + \frac{1}{2}(i\mathbf{k}\mathbf{r})^2 + \ldots \qquad (2.34)$$

where the successive terms in the expansion give the higher-order transitions (dipole, quadrupole, etc.).

If the electron is in a centrally symmetric potential, its wave function is $\Psi(r, \theta, \varphi) = R_{N,l}(r) Y_{l,m}(\theta, \varphi)$, where N, l, m are the three quantum numbers of the state and the spherical harmonic $Y_{l,m}(\theta, \varphi)$ is the angular-dependent part of the wave function.

Combining these results, the transition probability for scattering from electron state ψ_α to state ψ_β is proportional to the square of the following matrix element, $\int d\mathbf{r}\Psi_\beta^* \Psi_\alpha \cos\theta \exp(i\mathbf{k}\mathbf{r})$, which is proportional to

$$\int r^2 dr d\Omega Y_{l_\alpha, m_\alpha}(\theta, \varphi) Y_{l_\beta, m_\beta}(\theta, \varphi) Y_{1,0}(\theta, \varphi) \qquad (2.35)$$

$$\times \left[1 + i\mathbf{k}\mathbf{r} + \frac{1}{2}(i\mathbf{k}\mathbf{r})^2 + \ldots\right]$$

where $d\Omega$ denotes the integral over angles and we have replaced $\cos\theta$ by $Y_{1,0}(\theta,\varphi)$.

The unity in the parentheses gives the electric-dipole term. According to the well-known properties of the spherical harmonics, the angular integrals vanish unless $\Delta m = m_\alpha - m_\beta = 0$ and $\Delta l = l_\beta - l_\alpha = \pm1$, which gives the selection rules for an electric-dipole transition.

The direction of propagation of wave $\mathbf{k}$ is perpendicular to the polarization (z-direction). Therefore, let $\mathbf{k}$ be in the y-direction and $\mathbf{kr} \sim y$. The next term in the parenthesis, which is the quadrupole term, thus gives a factor yz in the integral, which can be written in terms of $Y_{2,-1}(\theta,\varphi)$ and $Y_{2,+1}(\theta,\varphi)$. The relevant angular integrals are $\int d\Omega Y_{l_\alpha,m_\alpha}(\theta,\varphi) Y_{l_\beta,m_\beta}(\theta,\varphi) Y_{2,-1}(\theta,\varphi)$ and $\int d\Omega Y_{l_\alpha,m_\alpha}(\theta,\varphi) Y_{l_\beta,m_\beta}(\theta,\varphi) Y_{2,+1}(\theta,\varphi)$. According to the properties of the spherical harmonics, these angular integrals vanish unless

$$\Delta m = m_\alpha - m_\beta = \pm1 \text{ and } \Delta l = l_\beta - l_\alpha = 0 \text{ or } \pm2 \qquad (2.36)$$

which (excluding the case $l_\beta = l_\alpha = 0$) gives the selection rules for an electric-quadrupole transition. These are the results for linear polarization. For arbitrary polarization, one obtains $\Delta m = 0$, or ±1 or ±2.

Magnetic-dipole transitions are determined by $\mathbf{L} = \mathbf{r} \times \mathbf{p}$. The relevant factor in the integrand is found to be L_x or L_y or L_z. Inserting these factors into the angular integrals $\int d\Omega \cos\theta Y_{l_\alpha,m_\alpha}(\theta,\varphi) (L_x \text{ or } L_y \text{ or } L_z) Y_{l_\beta,m_\beta}(\theta,\varphi)$ and using the properties of the spherical harmonics, yields the selection rules for a magnetic-dipole transition

$$\Delta m = m_\alpha - m_\beta = 0 \text{ or } \pm1 \text{ and } \Delta l = l_\beta - l_\alpha = \pm1 \qquad (2.37)$$

Problem 2.2.

Find the selection rules for an electron in an atom with strong spin-orbit interaction.

Solution.

One can easily find the selection rules for an electron in hydrogen-like atoms having only one electron in the unfilled shell for the case where the spin-orbit interaction is weak. Then, the atomic states are defined by the three quantum numbers, N, l, m, and the transition of an electron from one level to another one occurs without change in the spin. Therefore, one may neglect the spin, and the transitions are described in problem 2.1 above. However, when one considers heavy hydrogen-like atoms, spin-orbit coupling is important and leads to the replacement of the orbital quantum number l by the quantum number $j = l + s = l \pm 1/2$. The selection rules then become $j_2 - j_1 = 0, \pm1$ or $l_2 - l_1 = \pm1$.

Problem 2.3.

Find the probability of a dipole transition from state $2S_{1/2}$ to state $2P_{1/2}$ of the hydrogen atom. The measured energy difference between these two states (Lamb shift) is 4.4×10^{-6} eV.

Solution.

Fermi's Golden Rule for the transition probability W from initial state (i), to final state (f) is $W = (2\pi/\hbar) \langle V_{f,\,i} \rangle^2 \delta (E_f - E_i)$. This expression can be evaluated to give the probability for a one-photon dipole transition

$$W = \frac{4}{3} \frac{\omega^3}{\hbar c^3} |ea|^2 \tag{2.38}$$

Inserting $\hbar\omega = 4.4 \times 10^{-6}$ eV, $a \approx 10^{-8}$ cm, and $e \approx 5 \times 10^{-10}$ cgs units, yields $W \approx 3 \times 10^{-10}$ sec^{-1}. Thus, the lifetime $\tau = W^{-1} \approx 100$ years.

Chapter 3

Many-body Problem

3.1 Fermions and bosons

In classical mechanics, at any moment one can identify the particles in the system, and then follow the trajectories of each particle. However, in quantum mechanics, there are no trajectories, and after its identification, one cannot follow each particle to the next moment. The proper concept in quantum mechanics is "how many particles" and not "which particles". Assume that two identical particles are described by the wave function $\Psi_k(\xi_1, \xi_2)$, i.e., they are located in state k and described by the coordinates ξ_1 and ξ_2. Let us interchange the particles. Since only $|\Psi|^2$ has the physical meaning, the new wave function $\Psi_k(\xi_2, \xi_1)$ may differ from the original function only by a phase factor, i.e.,

$$\Psi_k(\xi_1, \xi_2) = \exp(i\alpha)\,\Psi_k(\xi_2, \xi_1) = \exp(2i\alpha)\,\Psi_k(\xi_1, \xi_2) \qquad (3.1)$$

In the last equality in (3.1), we return to the original function, which implies that $\exp(2i\alpha) = 1$ or $\exp(i\alpha) = \pm 1$. Therefore, all particles may be divided into two groups, bosons and fermions, according to

$$\Psi_k(\xi_1, \xi_2) = +\Psi_k(\xi_2, \xi_1) \quad \text{– bosons} \qquad (3.2)$$
$$\Psi_k(\xi_1, \xi_2) = -\Psi_k(\xi_2, \xi_1) \quad \text{– fermions}$$

Two fermions cannot be at the same state since $\Psi = 0$ for $\xi_1 = \xi_2$. This is the famous Pauli exclusion principle. If a system contains many fermions, the interchange of each pair of particles produces a minus sign. If the particles do not interact and thus can be described by single-particle wave functions $\Psi_k(\xi_j)$, a convenient formulation of the many-body wave function $\Psi(\xi_1, \xi_2...\xi_n)$ is a Slater determinant,

$$\Psi\left(\xi_1,\xi_2...\xi_n\right) = \begin{vmatrix} \Psi_{k_1}\left(\xi_1\right) & \Psi_{k_2}\left(\xi_1\right) & ... & \Psi_{k_n}\left(\xi_1\right) \\ \Psi_{k_1}\left(\xi_2\right) & \Psi_{k_2}\left(\xi_2\right) & ... & \Psi_{k_n}\left(\xi_2\right) \\ ... & ... & ... & ... \\ \Psi_{k_1}\left(\xi_n\right) & \Psi_{k_2}\left(\xi_n\right) & ... & \Psi_{k_n}\left(\xi_n\right) \end{vmatrix}, \tag{3.3}$$

The properties of a determinant guarantee that $\Psi\left(\xi_1,\xi_2...\xi_n\right)$ changes its sign upon interchanging two lines and vanishes whenever two lines are identical.

3.2 N-representation

One can define a system of identical particles by their distribution among different states, so that the wave function can be written in the form

$$|n_1, n_2...\rangle \tag{3.4}$$

where the numbers $n_1, n_2...$ can be arbitrary for bosons, but only $n_i = 0$ or $n_i = 1$ for fermions. One defines the creation operator a_k^+ and the destruction operator a_k for bosons, as the operators which increase or decrease by one the number of particles in the k state,

$$a_k\left|...n_k...\right\rangle = \sqrt{n_k}\left|...n_k - 1...\right\rangle; \tag{3.5}$$
$$a_k^+\left|...n_k...\right\rangle = \sqrt{n_k + 1}\left|...n_k + 1...\right\rangle$$

If there are no particles in the k state,

$$a_k\left|...0_k...\right\rangle = 0 \tag{3.6}$$

The eigenstates of the operators a_k and a_k^+ are chosen in such a way that the effect of a pair of these operators is the following,

$$a_k^+ a_k\left|...n_k...\right\rangle = \sqrt{n_k}a_k^+\left|...n_k - 1...\right\rangle = n_k\left|...n_k...\right\rangle \tag{3.7}$$
$$a_k a_k^+\left|...n_k...\right\rangle = \sqrt{n_k + 1}a_k\left|...n_k + 1..\right\rangle = (n_k + 1)\left|...n_k...\right\rangle$$

It follows that the commutation relations for the bosonic operators are

$$a_k a_k^+ - a_k^+ a_k = 1 \tag{3.8}$$

For the fermionic operators, one defines

$$a_k\left|...n_k...\right\rangle = n_k\left|...n_k - 1...\right\rangle; \tag{3.9}$$
$$a_k^+\left|...n_k...\right\rangle = (1 - n_k)\left|...n_k + 1...\right\rangle$$

According to the exclusion principle,

$$a_k\left|...0_k...\right\rangle = 0; \qquad a_k^+\left|...1_k...\right\rangle = 0 \tag{3.10}$$

The repeated action of these operators is given by

$$a_k^+ a_k \left|...n_k...\right\rangle = n_k a_k^+ \left|...n_k - 1...\right\rangle = n_k^2 \left|...n_k...\right\rangle$$

$$a_k a_k^+ \left|...n_k...\right\rangle = (1 - n_k) a_k \left|...n_k + 1...\right\rangle \qquad (3.11)$$

$$= (1 - n_k)(1 + n_k) \left|...n_k...\right\rangle$$

so that the commutation relations for the fermion operators are

$$a_k a_k^+ + a_k^+ a_k = \left(1 - n_k^2\right) + n_k^2 = 1; \qquad (3.12)$$

3.3 Lagrangian and Hamiltonian of quantum systems

As pointed out in Chapter 1, the Lagrangian of a system is chosen in such a way that Lagrange's equations agree with the equations of motion. In our case, the Schrödinger equation can be obtained from the following Lagrangian,

$$L = i\hbar \Psi^* \frac{d\Psi}{dt} - \frac{\hbar^2}{2m} \nabla \Psi^* \nabla \Psi - V \Psi^* \Psi \qquad (3.13)$$

with the momentum P given by the usual definition,

$$P = \frac{\partial L}{\partial (d\Psi/dt)} = i\hbar \Psi^* \qquad (3.14)$$

For coordinate Ψ, Lagrange's equation (1.42) reduces to

$$-V\Psi^* - i\hbar \frac{d\Psi^*}{dt} + \frac{\hbar^2}{2m} \nabla^2 \Psi^* = 0 \qquad (3.15)$$

while a coordinate Ψ^* leads to the Scrödinger equation for the function Ψ.

The Hamiltonian H, defined in Section 1.1.2, takes the following form

$$H = \int d\mathbf{r} \left(\frac{\hbar^2}{2m} \nabla \Psi^* \nabla \Psi + V \Psi^* \Psi \right) \qquad (3.16)$$

3.4 Second quantization

The Schrödinger equation for the wave function Ψ was obtained in Section 1.3.1 by replacing the physical quantities by operators. Second quantization means that the wave functions Ψ and Ψ^* in the Schrödinger equation are replaced by the following operators,

$$P_i(\mathbf{r}_1) \Psi_k(\mathbf{r}_2) - \Psi_k(\mathbf{r}_2) P_i(\mathbf{r}_1) = -i\hbar\delta(\mathbf{r}_1 - \mathbf{r}_2)\delta_{i,k};$$

$$\Psi_i(\mathbf{r}_1) \Psi_k(\mathbf{r}_2) = \Psi_k(\mathbf{r}_2) \Psi_i(\mathbf{r}_1) \qquad (3.17)$$

$$P_i(\mathbf{r}_1) P_k(\mathbf{r}_2) = P_k(\mathbf{r}_2) P_i(\mathbf{r}_1)$$

Note that Eq. (3.17) holds for the operators of the coordinate x and momentum $p_x = -i\hbar d/dx$ of the particle.

Inserting Eq. (3.14) into (3.17) leads to

$$\Psi_i(\mathbf{r}_1)\Psi_k^*(\mathbf{r}_2) - \Psi_k^*(\mathbf{r}_2)\Psi_i(\mathbf{r}_1) = \delta(\mathbf{r}_1 - \mathbf{r}_2)\delta_{i,k} \qquad (3.18)$$

We now study the properties of the operators Ψ and Ψ^*. Taking into account Eq. (3.18), their dynamic behavior is defined by Eq. (1.50),

$$\frac{\partial\Psi}{\partial t} \equiv \frac{i}{\hbar}(H\Psi - \Psi H) = \frac{i}{\hbar}\left(\frac{\hbar^2}{2m}\nabla^2\Psi - V\Psi\right) \qquad (3.19)$$

which gives the Schrödinger equation for operators. By replacing Ψ by H in Eq. (3.19), one sees that for a time-independent Hamiltonian, the energy is conserved. Another conserved quantity is the number of particles, given by the operator

$$N = \int \Psi^*(\mathbf{r})\Psi(\mathbf{r})\,d\mathbf{r} \qquad (3.20)$$

This result is confirmed by (3.18), if one replaces Ψ by N in Eq. (3.19).

3.5   Hamiltonian in second quantization

One turns from the operators $\Psi(\mathbf{r},t)$ and $\Psi^*(\mathbf{r},t)$ to the operators $a_k(t)$ and $a_k^+(t)$ by using an arbitrary set of functions $u_k(\mathbf{r})$, $u_k^*(\mathbf{r})$, defined by

$$\Psi(\mathbf{r},t) = \sum_k a_k(t)u_k(\mathbf{r}); \quad \Psi^*(\mathbf{r},t) = \sum_k a_k^+(t)u_k^*(\mathbf{r}) \qquad (3.21)$$

Comparing Eq. (3.18) with Eqs. (3.8) and (3.12) shows that the operators $a_k(t)$ and $a_k^+(t)$ are the destruction and creation operators introduced in the previous Section. The operator of the number of particles, introduced in Eq. (3.20), then takes the following form,

$$N = \sum_{k,l} a_k^+ a_l \int u_k^*(\mathbf{r})u_l(\mathbf{r})\,d\mathbf{r} = \sum_k a_k^+ a_k = \sum_k n_k \qquad (3.22)$$

Using the commutation relation for fermions and bosons, one can easily confirm the commutation relation of the operators $a_k^+ a_k$ and $a_l^+ a_l$ for the number of particles n_k and n_l in states k and l, $n_k n_l = n_l n_k$. Moreover, for fermions,

$$n_k^2 = a_k^+ a_k a_k^+ a_k = a_k^+ a_k(1 - a_k a_k^+) = a_k^+ a_k = n_k, \qquad (3.23)$$

which is consistent with $n_k = 0$ or $n_k = 1$, as obtained in the previous section.

The Hamiltonian

$$H = -\frac{\hbar^2}{2m}\nabla^2 + V(\mathbf{r}), \tag{3.24}$$

defined by the operators Ψ and Ψ^*, takes the following form

$$H = \int \Psi^* \left(-\frac{\hbar^2}{2m}\nabla^2 + V(\mathbf{r}) \right) \Psi dr \tag{3.25}$$

$$= \sum_{k,l} a_k^+ a_l \int u_k^* \left(-\frac{\hbar^2}{2m}\nabla^2 + V(\mathbf{r}) \right) u_l dr$$

By selecting plane waves $\exp(i\mathbf{kr})$ for the functions $u_k(\mathbf{r})$, one obtains for free particles ($V = 0$) of energy ε_k,

$$H_0 = \int \Psi^* \left(-\frac{\hbar^2}{2m}\nabla^2 \right) \Psi dr = \sum_k \varepsilon_k a_k^+ a_k, \tag{3.26}$$

and for the part of the Hamiltonian that describes the interaction between particles,

$$\int \Psi^* V(\mathbf{r_1}-\mathbf{r_2}) \Psi dr = \sum_{k,l,m,n} a_k^+ a_l^+ a_m a_n \tag{3.27}$$

$$\times \int \exp\left[-i(\mathbf{kr_1} + \mathbf{lr_2})\right] V(\mathbf{r_1}-\mathbf{r_2}) \exp\left[i(\mathbf{mr_1} + \mathbf{nr_2})\right] dr_1 dr_2$$

Changing variables from $\mathbf{r_1}$ and $\mathbf{r_2}$ to $\mathbf{r_1}$ and $\boldsymbol{\rho} = \mathbf{r_1} - \mathbf{r_2}$ reduces Eq. (3.27) to

$$\sum_{k,l,m,n} V_{klmn} a_k^+ a_l^+ a_m a_n \tag{3.28}$$

where

$$V_{klmn} = \delta_{k+l,\,m+n} \int V(\boldsymbol{\rho}) \exp\left[i(\mathbf{l} - \mathbf{n})\boldsymbol{\rho}\right] d\boldsymbol{\rho} \tag{3.29}$$

Combining (3.26) and (3.28), one obtains the Hamiltonian in second quantization,

$$H = \sum_k \varepsilon_k a_k^+ a_k + \sum_{k+l=m+n} V_{klmn} a_k^+ a_l^+ a_m a_n \tag{3.30}$$

In Section 3.8, we will use this Hamiltonian to analyze the phenomena of superfluidity and superconductivity.

3.6   Quantization of the electromagnetic field

In this section we shall show that by a series of transformations, one can describe an electromagnetic wave as a system of harmonic oscillators - bosonic particles (photons) with energy $\varepsilon_k = \hbar c k$ and momentum $\mathbf{p} = \hbar \mathbf{k}$. Consider the electromagnetic field in vacuum without charges and currents, which allows one to use the Coulomb gauge, $\nabla \mathbf{A} = 0$ and $\phi = 0$. The Lagrangian (1.43) and the Hamiltonian (1.45) then take the following forms

$$L = \frac{1}{8\pi c^2}\left(\frac{\partial \mathbf{A}}{\partial t}\right)^2 - \frac{1}{8\pi}\left(\nabla \times \mathbf{A}\right)^2 \tag{3.31}$$

$$H = \int \left[2\pi c^2 P^2 + \frac{1}{8\pi}\left(\nabla \times \mathbf{A}\right)^2\right] d\mathbf{r}$$

The generalized momentum $\mathbf{P}$, conjugate to the coordinate $\mathbf{A}$, is

$$\mathbf{P} \equiv \frac{\partial L}{\partial \left(\partial \mathbf{A}/\partial t\right)} = -\frac{\mathbf{E}}{4\pi c} \tag{3.32}$$

which gives for the electric $\mathbf{E}$ and magnetic $\mathbf{B}$ components of the field,

$$\mathbf{E} = -4\pi c \mathbf{P}; \qquad \mathbf{B} = \nabla \times \mathbf{A} \tag{3.33}$$

Second quantization of the coordinate $\mathbf{A}$ and momentum $\mathbf{P}$ of an electromagnetic field is performed analogously to the quantization of the Schrödinger equation carried out in Section 3.4. The operators $\mathbf{A}$ and $\mathbf{P}$ satisfy the following commutation relations

$$P_i\left(\mathbf{r}_1\right) A_k\left(\mathbf{r}_2\right) - A_k\left(\mathbf{r}_2\right) P_i\left(\mathbf{r}_1\right) = -i\hbar\delta\left(\mathbf{r}_1 - \mathbf{r}_2\right)\delta_{i,k} \tag{3.34}$$

$$A_i\left(\mathbf{r}_1\right) A_k\left(\mathbf{r}_2\right) = A_k\left(\mathbf{r}_2\right) A_i\left(\mathbf{r}_1\right)$$

$$P_i\left(\mathbf{r}_1\right) P_k\left(\mathbf{r}_2\right) = P_k\left(\mathbf{r}_2\right) P_i\left(\mathbf{r}_1\right)$$

Using Eq. (3.34), one can find the commutation relations for the components of the fields $\mathbf{E}$ and $\mathbf{B}$, defined in (3.33),

$$E_k\left(\mathbf{r}_1\right) E_k\left(\mathbf{r}_2\right) = E_k\left(\mathbf{r}_2\right) E_k\left(\mathbf{r}_1\right) \tag{3.35}$$

$$B_k\left(\mathbf{r}_1\right) B_k\left(\mathbf{r}_2\right) = B_k\left(\mathbf{r}_2\right) B_k\left(r_1\right)$$

$$E_k\left(\mathbf{r}_1\right) B_k\left(\mathbf{r}_2\right) = B_k\left(\mathbf{r}_2\right) E_k\left(\mathbf{r}_1\right)$$

where $k = x, y, z$ and

$$E_x\left(\mathbf{r}_1\right) B_y\left(\mathbf{r}_2\right) - B_y\left(\mathbf{r}_2\right) E_x\left(\mathbf{r}_1\right) = 4i\hbar\pi c\frac{\partial}{\partial z}\left[\delta\left(\mathbf{r}_1 - \mathbf{r}_2\right)\right] \tag{3.36}$$

with cyclic permutations of the indices x, y and z.

Analogous to the case of the Schrödinger equation, the dynamic equations for operators $\mathbf{E}$ and $\mathbf{B}$, defined in (1.50), coincide with Maxwell's equations (1.18)-(1.19), where $\mathbf{E}$ and $\mathbf{B}$ are replaced by the corresponding operators.

We will now replace the operators. First, instead of the operators $\mathbf{A}$ and $\mathbf{P}$, one introduces new operators $q_{k,\lambda}(t)$ and $p_{k,\lambda}(t)$, defined by

$$\mathbf{A}(\mathbf{r},t) = \sum_{k,\lambda} \left(q_{k,\lambda} u_{k,\lambda} + q_{k,\lambda}^+ u_{k,\lambda}^* \right), \tag{3.37}$$

$$\mathbf{P}(\mathbf{r},t) = \sum_{k,\lambda} \left(p_{k,\lambda} u_{k,\lambda} + p_{k,\lambda}^+ u_{k,\lambda}^* \right),$$

where the $u_{k,\lambda}(\mathbf{r})$ are plane waves and $\lambda = 1, 2$ is the polarization index. Equations (3.34) yield the commutation relations for the operators $q_{k,\lambda}(t)$ and $p_{k,\lambda}(t)$,

$$q_{k,\lambda_1} p_{l,\lambda_2}^+ - p_{l,\lambda_2}^+ q_{k,\lambda_1} = i\hbar \delta_{k,l} \delta_{\lambda_1,\lambda_2} \tag{3.38}$$

$$q_{l,\lambda_2}^+ p_{k,\lambda_1} - p q_{l,\lambda_2}^+ = i\hbar \delta_{k,l} \delta_{\lambda_1,\lambda_2}$$

while all other operators commute.

Turning now from the operators $\mathbf{A}$ and $\mathbf{P}$ in the Hamiltonian (3.31) to the operators $q_{k,\lambda}(t)$ and $p_{k,\lambda}(t)$, one obtains

$$H = \sum_{k,\lambda} \left(4\pi c^2 p_{k,\lambda} p_{k,\lambda}^+ + \frac{k^2}{4\pi} q_{k,\lambda} q_{k,\lambda}^+ \right) \tag{3.39}$$

The equations of motion of operators $p_{k,\lambda}$ and $q_{k,\lambda}$ follow from Eqs. (3.39) and (3.38),

$$\frac{dp_{k,\lambda}}{dt} = \frac{i}{\hbar} \left(Hp_{k,\lambda} - p_{k,\lambda} H \right) = -\frac{k^2}{4\pi} q_{k,\lambda} \tag{3.40}$$

$$\frac{dq_{k,\lambda}}{dt} = \frac{i}{\hbar} \left(Hq_{k,\lambda} - q_{k,\lambda} H \right) = 4\pi c^2 p_{k,\lambda}$$

Combining the two equations in (3.40) leads to equations for $p_{k,\lambda}$ and $q_{k,\lambda}$ having harmonic oscillator solutions of the form

$$q_{k,\lambda} = a_{k,\lambda} \exp(-ikct) + b_{k,\lambda}^+ \exp(ikct); \quad b_{k,\lambda}^+ = a_{-k,\lambda} \tag{3.41}$$

$$p_{k,\lambda} = \frac{1}{4\pi c^2} \frac{dq_{k,\lambda}}{dt} = \frac{ik}{4\pi c} \left[b_{k,\lambda}^+ \exp(ikct) - a_{k,\lambda} \exp(-ikct) \right]$$

Equations (3.41) permit one to express the operators $a_{k,\lambda}$ and $b_{k,\lambda}$ in terms of the operators $p_{k,\lambda}$ and $q_{k,\lambda}$. Using the commutation relations (3.38) gives

$$a_{k,\lambda_1} a_{l,\lambda_2}^+ - a_{l,\lambda_2}^+ a_{k,\lambda_1} = b_{k,\lambda_1} b_{l,\lambda_2}^+ - b_{l,\lambda_2}^+ b_{k,\lambda_1} = \frac{2\pi\hbar c}{k} \delta_{k,l} \delta_{\lambda_1 \lambda_2} \tag{3.42}$$

The final change of operators is

$$c_{k,\lambda} = \sqrt{\frac{k}{2\pi\hbar c}} a_{k,\lambda}; \quad d_{k,\lambda} = \sqrt{\frac{k}{2\pi\hbar c}} b_{k,\lambda} \tag{3.43}$$

with commutation relations (3.8) for bosonic particles,

$$c_{k,\lambda_1}c_{l,\lambda_2}^+ - c_{l,\lambda_2}^+ c_{k,\lambda_1} = d_{k,\lambda_1}d_{l,\lambda_2}^+ - d_{l,\lambda_2}^+ d_{k,\lambda_1} = \delta_{k,l}\delta_{\lambda_1\lambda_2} \qquad (3.44)$$

Inserting the operators $q_{k,\lambda}(t)$, $p_{k,\lambda}(t)$, $a_{k,\lambda}$, $b_{k,\lambda}$, $c_{k,\lambda}$, $d_{k,\lambda}$ into the Hamiltonian (3.39) leads to

$$H = \sum_{k,\lambda} \hbar c k \left(c_{k,\lambda}^+ c_{k,\lambda} + d_{k,\lambda}^+ d_{k,\lambda} \right) = \sum_{k,\lambda} \hbar \omega_k \left(N_{k,\lambda} + \frac{1}{2} \right) \qquad (3.45)$$

where $\omega_k = ck$. In the last equality in (3.45), the numbers of particles $c_{k,\lambda}^+ c_{k,\lambda}$ and $d_{k,\lambda}^+ d_{k,\lambda}$ with positive and negative wave vector $\mathbf{k}$ have been combined under the summation over all values of $\mathbf{k}$.

An analogous procedure can be performed for the momentum of an electromagnetic field. In the framework of second quantization, the general equation for $\mathbf{P}$ becomes an operator equation

$$\mathbf{P} = \frac{1}{4\pi c} \int (\mathbf{E} \times \mathbf{B}) \, d\mathbf{r} = - \int (\mathbf{P} \times \nabla \times \mathbf{A}) \, d\mathbf{r} \qquad (3.46)$$

According to equations (3.33), (3.37), (3.41) and (3.43), one can successfully transform $\mathbf{P}, \mathbf{A} \to p, q \to a, b \to c, d$. These transform Eq. (3.46) into the following form

$$\mathbf{P} = \sum_{k,\lambda.} \hbar \mathbf{k} N_{k,\lambda} \qquad (3.47)$$

Equations (3.45) and (3.47) complete our proof of the dual particle-wave nature of the electromagnetic field.

3.7 Full quantum mechanical description of the particle-wave interaction

In the discussion of the interaction of a charged particle with an electromagnetic field in Chapter 2, we considered a quantum particle interacting with a classical field. Here we consider both particle and field as quantum objects described by the Hamiltonian,

$$H = \sum_{k} \varepsilon_k n_k + \sum_{k} \hbar c k \left(N_k + \frac{1}{2} \right) \qquad (3.48)$$

As in Section 1.2.3, one uses perturbation theory, in which the potential $V = ie\hbar \mathbf{A}\nabla/mc$ is the perturbation, and the operator $\mathbf{A}$ is defined in Eq.

(3.37). Performing the transition from A to $c_{k,\lambda}$ and using Eq. (1.29) leads to

$$A = \sum_k \sqrt{\frac{2\pi\hbar c}{k}} \{c_{k,\lambda} \exp[i(\mathbf{kr} - \omega t)] + c.c.\} \qquad (3.49)$$

The first term in (3.49) contains the destruction operator $c_{k,\lambda}$, describing the disappearance of a photon, i.e., the absorption of a photon by the particle, whereas the creation operator $c_{k,\lambda}^+$ in the second term describes the emission of a photon by the particle. The probability of the transition of a system (particle and field) from one state (the particle in state n and the field having N_k photons in state k) to another state (the particle in state m and the field having $N_k \pm 1$ photons in state k) with absorption (emission) of a photon, is given by the matrix element,

$$V(n, N_k \rightarrow m, N_{k\pm1}) = \frac{ie\hbar}{mc} \int \Psi_{m,N_k\pm1}^* A \Psi_{n,N_k} dr \qquad (3.50)$$

where the wave function of the particle $u_k(r)\exp(i\omega_k t)$ is present in the wave function Ψ of the entire system. Progressing now to second quantization leads to

$$V = \frac{ie\hbar}{mc} \sum_{n,m} a_m^+ a_n \sqrt{\frac{2\pi\hbar c}{k}} \qquad (3.51)$$

$$\times \left\{ c_{k,\lambda} \exp[i(\omega_n - \omega_m - \omega)t] \int \exp(i\mathbf{kr}) u_n^*(\mathbf{r}) A_0 \nabla u_m(\mathbf{r}) dr + c.c. \right\}$$

As was shown in Section 2.1, the transition probability per unit time W is defined by the square of the time integral of (3.51). For the absorption of a photon, this gives

$$W = \frac{4\pi e^2 \hbar}{m^2 c\omega} N_k \left[\int \exp(i\mathbf{kr}) u_n^*(\mathbf{r}) A_0 \nabla u_m(\mathbf{r}) dr \right]^2 \qquad (3.52)$$

where the number of photons N_k can be expressed in terms of the intensity $I(\omega)$ of the field, $N_k = I(\omega)/\hbar ck$. For the emission of a photon, one has to replace N_k in Eq. (3.52) by $N_k + 1$. This means that (spontaneous) emission may occur even in the absence of an external field ($N_k = 0$). This effect cannot be described by the classical description of the field since this transition is connected with the polarization of the vacuum (to be discussed in Chapter 5).

This method permits one to obtain the Planck formula for black-body radiation. Consider the equilibrium between atoms and radiation based on the absorption-emission of photons. Assume that the probability of the transition from state k to state l, accompanied by the absorption of a photon, is proportional to the number N_k of particles in state k, while the probability of the inverse transition, accompanied by the emission of a photon, is proportional to $N_l + 1$, where the unity arises from spontaneous emission. Both induced transitions are proportional to the intensity $I(\omega)$ of the field. Therefore, in equilibrium one obtains,

$$B_{kl} I(\omega) N_k = B_{lk} I(\omega) N_l + A N_l \qquad (3.53)$$

where B_{ij} is the probability of the transition from state j to state i and A is the probability of spontaneous emission. Taking into account the conservation of energy $E_l - E_k = \hbar\omega$, and that the numbers of particles in states k and l depend on their energies, $N_k \approx \exp(-E_k/\kappa T)$, $N_l \approx \exp(-E_l/\kappa T)$, one obtains the Planck formula from Eq. (3.53)

$$I(\omega) = \frac{A/B}{\exp(\hbar\omega/\kappa T) - 1} \qquad (3.54)$$

3.8 Superfluidity and superconductivity

The Hamiltonian (3.30) of a system of particles allows one to understand the phenomena of superfluidity and superconductivity - the two most interesting phenomena in condensed matter physics. Superfluidity - the ability of a system of bosonic particles to flow without friction through any pipe - has the same physical explanation as superconductivity - the vanishing of the electrical resistance in a fermionic system. Indeed, superconductivity occurs because the fermion system of electrons in a solid forms boson pairs, and the superfluid motion of the charged paires leads to superconductivity.

The phenomenological explanation of superfluidity is based on the universally accepted belief that the low-energy excitations of a system can be considered as a set of "elementary excitations" - an ideal gas of quasiparticles with dispersion law $\varepsilon = \varepsilon(p)$. Let us assume that the liquid moves with velocity v with respect to the walls. Going to a new coordinate system in which the liquid is at rest and the walls are moving with velocity $-v$ with respect to the liquid, one can characterize the interaction between liquid and walls by the appearance in the liquid of excitations with energy ε and momentum $\mathbf{p}$. According to the Lorentz transformation, on returning to the original coordinate system, the change in the energy of the liquid will be $\varepsilon + \mathbf{p}v$. The minimal value of this quantity is $\varepsilon - pv$, when $\mathbf{p}$ is antiparallel to $\mathbf{v}$. Therefore, this process will be energetically favorable when

$\varepsilon - pv < 0$, that is, when $v > \varepsilon/p$. However, if $v < \varepsilon/p$, the transfer of the energy is not favorable, and the moving fluid will not interact with the walls, i.e., it will exhibit superfluidity. This brings us to the key question of the dispersion law $\varepsilon = \varepsilon(p)$ of the elementary excitations. If the minimum value of the ratio ε/p is zero, as is the case for a free particle for which $\varepsilon = p^2/2m$, then $v > 0$ always, and there is no superfluidity. If, however, the minimum value of $\varepsilon/p \neq 0$, as for the dispersion law $\varepsilon = ap$, then for velocity $v < a$, the liquid will exhibit superfluidity. The ground state and properties of the excitations can only be determined in the framework of a microscopic theory, which we shall now develop from the general properties of a system of bosons.

3.8.1 *Statistics of a bosonic gas*

Consider the statistical properties of a gas of non-interacting bosons with n_k particles in level k having energy ε_k. This system has energy E and number of particles N, given by

$$N = \sum_k n_k; \qquad E = \sum_k n_k \varepsilon_k \qquad (3.55)$$

The partition function Z and the grand canonical free energy Ω have the following form,

$$Z = \sum_N \exp\left(\frac{\mu N}{\kappa T}\right) \sum_n \exp\left(\frac{E_{n,N}}{\kappa T}\right); \quad \Omega = -\kappa T \ln Z, \qquad (3.56)$$

where the chemical potential μ is determined by the requirement that the number of particles remains unchanged. Inserting (3.55) into (3.56) gives for state k,

$$\Omega_k = -\kappa T \ln \sum_{n_k} \exp\left[\left(\frac{\mu - \varepsilon_k}{\kappa T}\right) n_k\right] \qquad (3.57)$$

$$= \kappa T \ln \left[1 - \exp\left(\frac{\mu - \varepsilon_k}{\kappa T}\right)\right]$$

The last equality in (3.57) is obtained by summing the geometric series. This series converges only if the common ratio is smaller than unity,

$$\exp\left(\frac{\mu - \varepsilon_k}{\kappa T}\right) < 1 \qquad (3.58)$$

for all ε_k including $\varepsilon_k = 0$. This requires that $\mu < 0$.

The average number of particles at level k, $\overline{n_k} = -\partial\Omega_k/\partial\mu$, is determined from Eq. (3.57),

$$\overline{n_k} = -\frac{\partial\Omega_k}{\partial\mu} = \frac{1}{\exp\left(\frac{\varepsilon_k-\mu}{\kappa T}\right) - 1}, \qquad (3.59)$$

which is precisely the Bose-Einstein distribution law. The conservation law for the number of particles

$$N = \sum_k \overline{n_k} = \sum_k \frac{1}{\exp\left(\frac{\varepsilon_k-\mu}{\kappa T}\right) - 1} \qquad (3.60)$$

is the equation for the chemical potential μ.

The summation over k in Eq. (3.60) can be replaced by an integration over $\varepsilon = \hbar^2 k^2/2m$,

$$N = \frac{gVm^{3/2}}{\sqrt{2}\pi^2\hbar^3} \int \frac{\sqrt{\varepsilon}d\varepsilon}{\exp\left(\frac{\varepsilon-\mu}{\kappa T}\right) - 1} \qquad (3.61)$$

$$= \frac{gV(m\kappa T)^{3/2}}{\sqrt{2}\pi^2\hbar^3} \int \frac{\sqrt{z}dz}{\exp\left(z - \frac{\mu}{\kappa T}\right) - 1}$$

where V is the volume of the system and $g = 2s+1$ is the Lande factor for the spin s.

Equation (3.61) gives the chemical potential μ as a function of the temperature and the particle density N/V. Something strange happens with this equation: μ is negative, and with decreasing of T, $|\mu|$ decreases, approaching zero at some temperature T_C. What will happen at temperatures below T_C? It is clear from Eq. (3.61) that our calculation neglects zero energy, $\varepsilon = 0$. Matters can be improved by the assumption that at $T = T_C$, some particles are in the state of zero energy (Bose-Einstein condensate), so that at $T < T_C$, the system consists of the condensate plus the particles with positive energies. For T just a bit below T_C, there are only a few particles in the condensate, whereas for T near zero, the system is mostly condensate with just a few excitations. In the following Section, we will present the microscopic analysis of the properties of these excitations.

3.8.2   *Microscopic theory of superfluidity*

The goal of the analysis is to introduce some approximations which will bring the Hamiltonian (3.30),

$$H = \sum_k \varepsilon_k a_k^+ a_k + \sum_{k+l=m+n} V_{klmn} a_k^+ a_l^+ a_m a_n \qquad (3.62)$$

into diagonal form,

$$H = \sum_k E(k) b_k^+ b_k \qquad (3.63)$$

where $E = E(k)$ is the dispersion law of the elementary excitations. Our assumption is that at low temperatures, $T \ll T_C$, almost all particles are in the condensate and have zero energy

$$n_0 \equiv a_0^+ a_0 \approx N \qquad (3.64)$$

while the number $n_k \equiv a_k^+ a_k$ of excitations with $k \neq 0$ is very small. These excitations may appear only in pairs due to the conservation law $k + l = m + n$ of Eq. (3.62). There are only a few possibilities, which we incorporate by the following approximate form of the Hamiltonian (3.62),

$$H = V(0) N^2 + \sum_k \varepsilon_k a_k^+ a_k + \qquad (3.65)$$

$$+ \sum_k V(k) \left[a_0^+ a_0^+ a_k a_{-k} + a_k^+ a_{-k}^+ a_0 a_0 + a_0^+ a_k^+ a_k a_0 + a_k^+ a_0^+ a_0 a_k \right] =$$

$$= V(0) N^2 + \sum_k \left[\varepsilon_k + 2N_0 V(k) \right] a_k^+ a_k + N_0 V(k) \left(a_k^+ a_{-k}^+ + a_k a_{-k} \right)$$

Introduce new operators b_k and b_k^+, defined by

$$b_k = u_k a_k + v_k a_{-k}^+; \qquad b_k^+ = u_k a_k^+ + v_k a_{-k} \qquad (3.66)$$

where the functions u_k and v_k satisfy the requirements of bosonic statistics. This reduces the Hamiltonian from the form (3.62) to (3.63). Using the bosonic commutation of the operators a_k and a_k^+, one can confirm from Eq. (3.66) the bosonic commutation for operators b_k and b_k^+, $b_k b_k^+ - b_k^+ b_k = 1$, if

$$u_k^2 - v_k^2 = 1 \qquad (3.67)$$

From Eq. (3.66), one obtains the operators a_k and a_k^+ as functions of the operators b_k and b_k^+, and inserts them into Eq. (3.65), which yields

$$H = H_0 + \qquad (3.68)$$

$$\sum_k \left\{ \left[\varepsilon_k + N_0 V(k) \right] \left(u_k^2 + v_k^2 \right) + 2V(k) u_k v_k \right\} \left(b_k^+ b_k + b_{-k}^+ b_{-k} \right) +$$

$$\sum_k \left\{ \left[2 \left(\varepsilon_k + N_0 V(k) \right) \right] u_k v_k + V(k) \left(u_k^2 + v_k^2 \right) \right\} \left(b_k^+ b_{-k}^+ + b b_{-k} \right)$$

The functions u_k and v_k are chosen to make the non-diagonal terms in Eq. (3.68) vanish. The Hamiltonian (3.68) then reduces to the form (3.63) with

$$E(k) = \sqrt{2N V(k) \varepsilon_k + \varepsilon_k^2} \qquad (3.69)$$

In the limiting case of large k, Eq. (3.69) reduces to the energy of a free particle, $E(k) = \varepsilon_k$, whereas for small k, the energy of the elementary excitations ("quasiparticles") is a linear function of momentum. The latter are just the excitations needed to explain the phenomenon of superfluidity.

3.8.3   *Experimental detection of superfluidity*

The theory of superfluidity given in the last section applies to ideal systems, and it was not clear whether the interactions between particles, which were ignored in the analysis but always exist in real systems, would destroy the effect of superfluidity. Another question is the regimes of density and temperature of the bosonic systems where this effect may occur - the unique phase transition in a system of non-interacting particles. A single characteristic temperature T_C can be formed from the parameters of an ideal quantum system in the form $T_C = \hbar^2 n^{2/3}/m\kappa$, which, for the parameters of helium, gives $T_C = 3.13$ K. There are contradictory demands on the density of a system, which has to be large enough for quantum effects to show up, but small enough for the quantum effects to show up before the system solidifies. In fact, superfluidity occurs at densities $10^{13} - 10^{15}$ cm^{-3} - much smaller than the density of air, 10^{19} cm^{-3}, or of solids, 10^{22} cm^{-3}. The manifestation of quantum effects requires that the wavelength λ of a particle be larger than the average distance a between particles, $\lambda > a$. The latter inequality can be rewritten as

$$\lambda = \frac{\hbar}{p} = \frac{\hbar}{mv} = \frac{\hbar}{m\sqrt{\kappa T/m}} = \frac{\hbar}{\sqrt{\kappa Tm}} > a \approx n^{-1/3}, \tag{3.70}$$

i.e., the mass of a particle m and the temperature T have to be small, but the density n not too small. Hydrogen is the lightest atom, but two hydrogen atoms with opposite spins link to form a hydrogen molecule. The next candidate is the bosonic isotope of helium He$_4$, where the effect of superfluidity was found experimentally in 1938 at $T_C = 2.17$ K, which is close to the theoretical prediction $T_C = 3.13$ K.

It remains to show why the Coulomb attraction does not destroy the electron pairing. The answer is provided by the following estimates of the characteristic distances ξ between electrons in the pair,

$$\xi \approx \frac{\hbar}{\delta p} = \frac{\hbar p_f^2}{\delta p \, p_f^2} = \frac{\hbar}{p_f} \frac{p_f^2}{\delta \varepsilon_f} = \frac{\hbar}{p_f} \frac{p_f^2}{\hbar \omega}, \tag{3.71}$$

which is much larger than the typical distance $\hbar/p_f$ between electrons. The situation is similar to two dancers which remain a pair even while being far apart.

More recent attempts to find superfluidity in heavier bosonic systems necessitate the availability of much lower temperatures. Sufficient low temperatures have been achieved by laser traps, where three perpendicular laser beams are able to stop a particle, and the magnetic (evaporative) trap, where an inhomogeneous magnetic field creates a potential barrier which allows hot atoms to climb over the barrier and leave the system. Such techniques allows one to "stop" atoms, reducing their velocity from 40.000 cm/sec at 300 K to 1 cm/sec at 10^{-8} K. The Bose-Einstein condensation and the superfluid phenomenon have been found experimentally in systems of *Rb*, *Cs*, *Na*, *Li* atoms, as well as in different mixtures, including mixtures of bosons and fermions. The phase transition temperatures in these systems are much lower than in helium although the percentage of atoms in the condensate is much higher. There is now clear experimental evidence of these phenomena, such as the shift of the maxima of average velocities, interference between two Bose-Einstein condensates, etc. Moreover, the condensate in *Rb* is only one fifth of the thickness of a sheet of paper, but one can see this quantum object in a video camera! The Bose-Einstein condensate, which is a super-close packed system that behaves as a giant single atom (like soldiers at a gala march), possesses many strange properties, such as "stopping" light by reducing its velocity to 17 m/sec! It is not surprising that this phenomenon has generated considerable excitement among scientists and the general public, who call it "superatom", "atomic laser", the "fifth state of matter", "coldest atom in the universe", and finally, "Einstein was right!".

3.8.4 *Electron pairing in superconductors*

As indicated above, a boson system exhibits the phenomenon of superfluidity, which for charged particles means superconductivity, i.e., the motion of an electric current without resistance. The combination of two electrons into an electron pair requires an attractive force to overcome the electron repulsion. As was shown by Cooper in 1956, even an infinitely small attraction between two electrons with opposite wave vectors located above the Fermi surface, leads to pairing. Consider the Schrödinger equation for two electrons,

$$-\frac{\hbar^2}{2m}\left(\nabla_1^2 + \nabla_2^2\right)\Psi\left(r_1, r_2\right) + V\left(r_1 - r_2\right)\Psi\left(r_1, r_2\right) \qquad (3.72)$$
$$= \left(E + 2\varepsilon_f\right)\Psi\left(r_1, r_2\right)$$

where the energy E is measured from the Fermi energy ε_f. Thus, $E < 0$ corresponds to the creation of an electron pair.

For two electrons having wave vectors $\pm k$, one can rewrite Eq. (3.72) in k space for the function g_k, defined by

$$\Psi\left(r_1, r_2\right) = \sum_k g_k \exp\left[ik\left(r_1 - r_2\right)\right] \tag{3.73}$$

which gives

$$\frac{\hbar^2 k^2}{m} g_k + \sum_{k_1} V\left(k - k_1\right) g_{k_1} = \left(E + 2\varepsilon_f\right) g_k \tag{3.74}$$

Assume that the interaction between electrons $V\left(k - k_1\right)$ is negative (attractive), $V\left(k - k_1\right) = -V_0$ in the small region between $\varepsilon_F - \hbar\omega$ and $\varepsilon_F + \hbar\omega$ around the Fermi energy. Solving Eq. (3.74), one obtains

$$g_k = \frac{V_0 \displaystyle\sum_{k_1} g_{k_1}}{\hbar^2 k^2/m - E - 2\varepsilon_f} \tag{3.75}$$

The summation of (3.75) over k yields

$$1 = \sum_k \frac{V_0}{\left(\hbar^2 k^2/m\right) - E - 2\varepsilon_f} \tag{3.76}$$

Transforming the summation over k into an integral with density of state $\rho\left(\varepsilon\right)$, and changing the variable $\hbar^2 k^2/m$ to $\left(\hbar^2 k^2/2m\right) - 2\varepsilon_f$, gives

$$E = -2\hbar\omega \exp\left(-\frac{2}{V_0 \rho\left(\varepsilon_f\right)}\right) \tag{3.77}$$

As expected, the energy is negative, which means attraction and the creation of an electron pair. It is remarkable that due to the exponential dependence in Eq. (3.77), this result cannot be obtained by perturbation theory no matter how small V_0 is. The effective attraction between electrons results from the oscillations of ions in the solid, just as the ball falling on a drum creates a depression which attracts a neighboring ball. The characteristic energy $\hbar\omega$ of the oscillating ions is about several hundred degrees, a value much smaller than ε_F, which is of order of tens of thousands of degrees. Equation (3.77) explains two other experimental facts: different critical temperatures for different isotopes of mass M ($\hbar\omega \propto M^{-1/2}$), and the destruction of superconductivity by an external force with energy larger than $|E|$.

3.8.5 *Microscopic theory of superconductivity*

After demonstrating the coupling of each two electrons into an electron
pair, the question arises of whether the interactions between pairs will still
conserve the gap in the electron spectrum near the Fermi surface, which
is responsible for the phenomenon of superconductivity. The calculation
here is quite similar to that performed in Section 3.8.2. Starting from the
Hamiltonian of a system of electron pairs, one introduces the operators of
the excitations, confirms their statistics, and, by diagonalizing the Hamiltonian, finds their excitation spectrum showing that this spectrum has a
gap. The Hamiltonian of the electron pairs follows immediately from Eq.
(3.62),

$$H = \sum_k \varepsilon_k c_k^+ c_k + \sum_{k_1} V_{k,\,k_1} c_{k_1}^+ c_{-k_1}^+ c_{-k} c_k \qquad (3.78)$$

where c_k and c_k^+ are the fermion operators. The operators $c_{-k}c_k$ and
$c_{k_1}^+ c_{-k_1}^+$ describe the destruction and creation of the electron pairs while
$V_{k,\,k_1}$ is the interaction between the pairs. One introduces the new operators d_k and d_k^+, defined by

$$c_k = u_k d_k + v_k d_{-k}^+; \qquad c_{-k} = u_k d_{-k} - v_k d_k^+ \qquad (3.79)$$

where the functions u_k and v_k satisfy the requirements of fermionic statistics
and reduce the Hamiltonian (3.78) to diagonal form. Using the fermionic
commutation relations between operators c_k and c_k^+, one can confirm from
Eq. (3.79) the bosonic commutation relations for the operators d_k and d_k^+,
$d_k d_k^+ - d_k^+ d_k = 1$, if one chooses u_k and v_k to satisfy

$$u_k^2 + v_k^2 = 1 \qquad (3.80)$$

Inserting Eq. (3.79) into Eq. (3.78) gives

$$H = H_0 + \qquad (3.81)$$

$$\sum_k \left(\varepsilon_k \left(u_k^2 - v_k^2 \right) + u_k v_k \sum_{k_1} V_{k,\,k_1} u_{k_1} v_{k_1} \right) \left(d_k^+ d_k + d_{-k}^+ d_{-k} \right) +$$

$$\sum_k \left(\varepsilon_k u_k v_k - \left(u_k^2 - v_k^2 \right) \sum_{k_1} V_{k,\,k_1} u_{k_1} v_{k_1} \right) \left(d_k^+ d_{-k}^+ + d_k d_{-k} \right)$$

The non-diagonal terms in (3.81) vanish by an appropriate choice of u_k
and v_k. Thus, the Hamiltonian reduces to diagonal form with the energy
of the excitations being

$$E(k) = \frac{\hbar^2}{2m} \left[(k - k_f)^2 + \left(\frac{2m\Delta_k}{\hbar^2} \right)^2 \right]^{1/2} \qquad (3.82)$$

with a gap Δ_k in the energy spectrum, $\Delta_k = \sum_{k_1} V_{k, \, k_1} u_{k_1} v_{k_1}$.

3.8.6   *High-temperature superconductivity*

Superconductivity was discovered a hundred years ago. The practical applications of this interesting phenomenon was restricted by the small magnitude of the critical current that destroys superconductivity and, primarily, by the requirement of very low temperatures. During the seventy years after this discovery, an active search for high-temperature superconductors produced only quite modest results, never reaching values exceeding 30 K. The breakthrough occurred in 1986 when Bednorz and Müller [11] discovered high-temperature superconductivity in cuprates whose critical temperature in some cuprates was higher than 100 K. Within a few years, many copper oxide (cuprates) high-temperature superconductors were synthesized. The highest critical temperature was reported for thallium-doped mercury, based on a cuprate with a critical temperature of 138 K, which is much higher than the temperature (77 K) of liquid nitrogen used for cooling. The great enthusiasm culminated in the Nobel prize being awarded to Bednorz and Müller only one year (!) after the publication of their results. The original system $La_{2-x}Ba_x CuO_4$ is highly anisotropic and exhibits superconductivity in two-dimensional Cu-O planes. However, the cuprates are ceramics, which hinders their technological use. Therefore, the search for high-temperature superconductors continued in different directions.

A layered structure, similar to cuprates, has been found in a non-cuprate group of superconductors called iron-pnictide compounds (compounds of the nitrogen group) with critical temperatures up to 50 K. As in the cuprates, this material is antiferromagnet at low doping, but increased doping destroys the antiferromagnetism, leading to superconductivity. Research on iron-based superconductors started in 2006, with $T_C = 4$ K in $LaOFeB$ [12], and since then, these materials have been studied intensively [13]. The nature of the pairing mechanism in the iron-pnictide superconductors is still unknown. However, similarities to the cuprate high-temperature superconductors suggest that a similar mechanism may be at work, although the electrons in these materials might be paired with the aid of spin fluctuations.

The new high-temperature superconductors are based on compounds centered around "fullerene", drawing its name from the designer-author of the geodesic dome on top of Buckingham Palace. The fullerene ("buckyball") consists of a sixty carbon atoms joined in a roughly spherical shape. When doped with alkali metals, the fullerene (named "fulleride") becomes a superconductor with T_C ranging from 8 K (with sodium) to 70 K when doped with the interhalogen compound *ICl*. Another application of the carbon atoms for high-temperature superconductivity was suggested recently [14], using the hydrocarbons in the form of picene ($C_{22}H_{14}$) doped with atoms of an alkali (rubidium, potassium, cesium or sodium), which becomes superconducting with T_C of 18 K. Fullerenes are technically part of a larger family of superconductors containing organic materials. The great advantage of these systems is their high resistance to breakdown of the superconductivity in the presence of an applied magnetic field as high as 6 Tesla, whereas fields a fraction of this strength would destroy most other superconductors.

Superconductivity was found in some systems which at first glance would not seem to allow it. It has always been assumed that superconductors cannot be formed from ferromagnetic transition metals such as iron, cobalt or nickel. However, in the "borocarbides", as in the fullerenes, the crystallographic sites for the magnetic ions are thought to be isolated from the conduction path. Another new class of high-temperature superconductors are the heavy fermions - compounds containing rare-earth elements such as Ce or Yb, or actinide elements such as uranium [15]. Their (inner shell) conduction electrons often have effective masses (known as quasiparticle masses) several hundred times as great as that of "normal" electrons, resulting in a low "Fermi energy". It seems likely that the Cooper pairing in the heavy fermion systems arises from the magnetic interactions of the electron spins (*s*-, *p*-, *d*- waves), rather than from lattice vibrations. Heavy fermion superconductivity has since been observed in $CeCu_2Si_2$ and in more than 20 other cerium compounds.

It is worthwhile mentioning two other materials which exhibit superconductivity. One is ruthenium-oxygen, Sr_2RuO_4, where the Ru-O planes play a role similar to the copper-oxygen planes in cuprates. Although the critical temperature in this compound is quite low [16], its existence opens a new area of research. In 1999, a new compound, ruthenium-cuprate, was found to have a critical temperature of 58 K [17].

Considerable excitement was generated in the physics community by the discovery of superconductivity with a transition temperature of 39 K in magnesium diboride MgB_2 [18], a well-known compound [19] with a structure quite different from that of other superconductors. This material is cheap and easy fabricate into superconducting wires [20].

The theory still faces the serious problem of explaining a large body of experimental results, partially described above. The main problem relates to the mechanism for electrons pairing. Is the electron-phonon interaction fundamental in spite of the restriction on its validity [21]?

3.9   Problems

Problem 3.1.

Find the first-order correction to the energy of the ground state of a many-body system ($N \gg 1$) of bosons and fermions perturbed by a short-range interaction potential between the particles. Use the second quantization representation.

Solution.

The interaction potential between the particles can be written as

$$U = \frac{1}{2} \sum_{k,l,m,n} V_{klmn} a_k^+ a_l^+ a_m a_n \tag{3.83}$$

where

$$V_{klmn} = \int V(\mathbf{r}_1 - \mathbf{r}_2) \exp\left[i(\mathbf{k} - \mathbf{m})\mathbf{r}_1 + i(\mathbf{l} - \mathbf{n})\mathbf{r}_2\right] d\mathbf{r}_1 d\mathbf{r}_2 \tag{3.84}$$

According to first-order of perturbation theory, the energy shift is given by $\Delta E = \langle \Psi_0 | U | \Psi_0 \rangle$, where Ψ_0 is the wave function of the non-interacting particles.

First consider a system of bosons, for which all particles in the ground state have zero momentum, $\mathbf{p} = \hbar \mathbf{k} = 0$. Therefore, the matrix element of the operator (3.84) is different from zero only for $k = l = m = n = 0$, and the number of terms contributing to the sum in (3.83) is equal to $N(N-1) \approx N^2$. Thus, the shift of the energy is $\Delta E \approx \frac{1}{2} V_{0000} N^2$. The integral in (3.84) for a short-range interaction potential $V(\mathbf{r})$ contains

$$\int V(\mathbf{r}_1 - \mathbf{r}_2) d\mathbf{r}_2 \approx \int V(\rho) d\rho = \langle V(0) \rangle \tag{3.85}$$

yielding

$$\Delta E \approx \frac{1}{2} \langle V(0) \rangle N^2 \tag{3.86}$$

For an ideal gas of fermions in the ground state, all energy levels up to the Fermi energy ε_F are occupied, and all indices in (3.83) correspond to states having energy smaller than ε_F. The conservation of momentum implies that $\mathbf{k} + \mathbf{l} = \mathbf{m} + \mathbf{n}$, which for fermions reduces to $\mathbf{k} = \mathbf{l}$ and $\mathbf{m} = \mathbf{n}$

---

or $\mathbf{k} = \mathbf{m}$ and $\mathbf{l} = \mathbf{n}$. Taking into account the spin variable σ, one obtains for the energy shift due to interactions

$$\Delta E = \frac{1}{2} \langle V(0) \rangle \times \tag{3.87}$$

$$\left\langle \Psi_0 \left| \sum_{k,l,\sigma_1,\sigma_2} \left(a^+_{k,\sigma_1} a^+_{l,\sigma_2} a_{l,\sigma_2} a_{k,\sigma_1} + a^+_{l,\sigma_1} a^+_{k,\sigma_2} a_{l,\sigma_2} a_{k,\sigma_1} \right) \right| \Psi_0 \right\rangle$$

Inserting into (3.87) the number operator $n_{k,\sigma} = a^+_{k,\sigma} a_{k,\sigma}$ with $\sum_k n_{k,\sigma} = N/2$, and keeping in (3.87) only the large first term, which reduces to $\sum n_{k,\sigma_1} n_{l,\sigma_2} = N^2/4$, one obtains Eq. (3.86) with $1/2$ replaced by $1/4$.

Problem 3.2.

Consider a one-dimensional infinite chain of bosons described by the Hamiltonian

$$H = \sum_j \left[J_1 \left(a^+_{j+1} a_j + a_{j+1} a^+_j \right) + J_2 \left(a^+_{j+1} a^+_j + a_j a_{j+1} \right) \right] \tag{3.88}$$

a) Fourier transform the operators a_j and a^+_j to a_q and a^+_q, and then replace the operators a_q and a^+_q by $a_q \exp(iq/2)$ and $a^+_q \exp - (iq/2)$.

b) Diagonalize the Hamiltonian by the following transformation, $b_k = u_k a_k + v_k a^+_k$, and find the corresponding functions u_k and v_k. Then, calculate the resulting energy of the quasiparticles.

Solution.

a) Replacing the operators a_j and a_{j+1} by their Fourier transforms,

$$a_j = \sum_q a_q \exp(iqj); \qquad a^+_j = \sum_q a^+_q \exp(-iqj) \tag{3.89}$$

the Hamiltonian (3.89) takes the following form,

$$H = \sum_j \{ J_1 \left[a^+_q a_q \exp(-iq) + a_q a^+_q \exp(iq) \right] + \tag{3.90}$$

$$J_2 \left[a^+_q a^+_{-q} \exp(-iq) + a_q a_{-q} \exp(iq) \right] \}$$

A further transformation $a_q \rightarrow a_q \exp(iq/2)$ and $a^+_q \rightarrow a^+_q \exp(-iq/2)$ transforms (3.90) into

$$H = \sum_j \left[J_1 \left(a^+_q a_q 2 \cos q \right) + J_2 \left(a^+_q a^+_{-q} + a_q a_{-q} \right) \right] + \sum_{q>0} 2 \cos q \tag{3.91}$$

b) Analogous to the calculations performed in Section 3.8.2, diagonalize the Hamiltonian (3.91) with the help of the Bogolyubov transformation, which gives for the energies of the quasiparticles

$$E_{q>0} = \frac{2}{J_1 \cos q} \left(J_1^2 u_q^2 \cos^2 q + J_2^2 \right),$$ (3.92)

$$E_{q<0} = \frac{2}{J_1 \cos q} \left(J_1^2 v_q^2 \cos^2 q - J_2^2 \right)$$

Chapter 4

S-matrix, Green Function, Feynman Diagrams

4.1 S-matrix

The S-matrix, $S\left(t_2 - t_1\right)$, defines the dynamics of a quantum system connecting the Ψ-function of a system at time t_2 with that at a previous time t_1. It is convenient to consider the S-matrix in the interaction representation given in Section 1.3.4. In the interaction representation, the Schrödinger equation has the following form,

$$i\hbar \frac{\partial \Psi_I}{\partial t} = H_I\left(t\right) \Psi_I \tag{4.1}$$

where $H_I\left(t\right) = \exp\left(iH_0t/\hbar\right) H_1 \exp\left(-iH_0t/\hbar\right)$ is given in terms of the components H_0 and H_1 of the Hamiltonian $H = H_0 + H_1$, which describes the unperturbate energies of the particles and their interactions, respectively. The integration of Eq. (4.1) converts the differential equation into an integral equation

$$\Psi_I\left(t\right) = \Psi_I\left(t_0\right) - \frac{i}{\hbar} \int_{t_0}^{t} H_I\left(t_1\right) \Psi_I\left(t_1\right) dt_1 \tag{4.2}$$

Using the method of successive approximations, one inserts into the integral the solution of the zeroth-order approximation $\Psi_I^{(0)}\left(t_1\right)$,

$$\Psi_I^{(1)}\left(t\right) = -\frac{i}{\hbar} \int_{t_0}^{t} H_I\left(t_1\right) \Psi_I^{(0)}\left(t_1\right) dt_1 \tag{4.3}$$

The second approximation becomes

$$\Psi_I^{(2)}(t) = -\frac{i}{\hbar} \int_{t_0}^{t} H_I(t_1) \Psi_I^{(1)}(t_1) \, dt_1 \tag{4.4}$$

$$= \left(-\frac{i}{\hbar}\right)^2 \int_{t_0}^{t} H_I(t_1) \, dt_1 \int_{t_0}^{t_1} H_I(t_2) \Psi_I^{(0)}(t_2) \, dt_2$$

One can simplify the integral in the right-hand side,

$$\int_{t_0}^{t} dt_1 \int_{t_0}^{t_1} dt_2 ... = \frac{1}{2} \int_{t_0}^{t} dt_1 \int_{t_0}^{t} dt_2 ... \tag{4.5}$$

Extending this process to higher approximations leads to

$$\Psi_I(t) = S(t - t_0) \Psi_I^{(0)}(t) \tag{4.6}$$

with the scattering matrix (S-matrix) given by

$$S(t - t_0) = 1 - \frac{i}{\hbar} \int_{t_0}^{t} H_I(t_1) \, dt_1$$

$$+ \frac{1}{2} \left(-\frac{i}{\hbar}\right)^2 \int_{t_0}^{t} dt_1 \int_{t_0}^{t} H_I(t_1) H_I(t_2) \, dt_2 + ... \tag{4.7}$$

$$+ \frac{1}{n!} \left(-\frac{i}{\hbar}\right)^n \int_{t_0}^{t} ... \int_{t_0}^{t} H_I(t_1) ... H_I(t_n) \, dt_1 ... dt_n + ...$$

Equation (4.7) can be rewritten as

$$S(t - t_0) = \hat{P} \exp\left(-\frac{i}{\hbar} \int_{t_0}^{t} H_I(t_1) \, dt_1\right) \tag{4.8}$$

where the chronological operator $\hat{P}$ ensures that each term of the expansion of Eq. (4.8) has the times $t_1, t_2 ... t_n$ in descending order from left to right.

4.2   Green function

The main use of the Green function $G(\mathbf{r}, t)$ in mathematics is to solve differential equations. The solution of the differential equation

$\hat{L}\Psi(\mathbf{r}, t) = f(\mathbf{r}, t)$ with a differential operator $\hat{L}$ is expressed in terms of the Green function as follows,

$$\Psi(\mathbf{r}, t) = \Psi_0(\mathbf{r}, t) + \int G(\mathbf{r} - \mathbf{r}_1, t - t_1) f(\mathbf{r}_1, t_1) d\mathbf{r}_1 dt_1 \qquad (4.9)$$

where

$$\hat{L}G(\mathbf{r} - \mathbf{r}_1, t - t_1) = \delta(\mathbf{r} - \mathbf{r}_1) \delta(t - t_1) \qquad (4.10)$$

One can easily verify Eq. (4.9) by applying the operator $\hat{L}$ to this equation and using (4.10). As an example, consider the Schrödinger equation for a particle in a potential V,

$$\left(i\hbar \frac{\partial}{\partial t} + \frac{\hbar^2}{2m} \nabla^2 - V(\mathbf{r}) \right) \Psi = 0 \qquad (4.11)$$

For a free particle ($V = 0$), the equation for the Green function G_0 has the following form

$$\left(i\hbar \frac{\partial}{\partial t} + \frac{\hbar^2}{2m} \nabla^2 \right) G_0(\mathbf{r} - \mathbf{r}_1, t - t_1) = \delta(\mathbf{r} - \mathbf{r}_1) \delta(t - t_1) \qquad (4.12)$$

Performing a space Fourier transform yields

$$\left(i\hbar \frac{\partial}{\partial t} - \frac{\hbar^2 k^2}{2m} \right) G_0(\mathbf{k}, t - t_1) = \delta(t - t_1) \qquad (4.13)$$

whose solution is

$$G_0(\mathbf{k}, t - t_1) = -\frac{i}{\hbar} \Theta(t - t_1) \exp\left[-\frac{i\varepsilon_k}{\hbar}(t - t_1) \right] \qquad (4.14)$$

where the Heaviside function $\Theta(t - t_1) = 0$ for $t < t_1$ and $= 1$ for $t > t_1$. Thus, $d\Theta(t - t_1)/dt = \delta(t - t_1)$.

From equation (4.11) with $V(\mathbf{r}) \neq 0$, one obtains

$$\hat{L}G(\mathbf{k}, t - t_1) \equiv \left(i\hbar \frac{\partial}{\partial t} - \frac{\hbar^2 k^2}{2m} - V(\mathbf{k}) \right) G(\mathbf{k}, t - t_1) = \delta(t - t_1) \qquad (4.15)$$

One cannot solve Eq. (4.15), but this equation can be rewritten in the form of an integral equation,

$$G(\mathbf{k}, t - t_1) = G_0(\mathbf{k}, t - t_1) + \int dt_2 G_0(\mathbf{k}, t - t_2) G(\mathbf{k}, t_2 - t_1) \qquad (4.16)$$

The equivalence of Eqs. (4.15) and (4.16) is confirmed by applying the operator $\hat{L}$ to (4.16). The latter equation is convenient for use in perturbation theory.

The Green function $G(\mathbf{k}_1, \mathbf{k}_2, t - t_1)$ has the physical meaning of a "propagator" - the probability amplitude that a particle starting in state

$\mathbf{k}_1$ at time t_1 will reach state $\mathbf{k}_2$ at time t_2. Since there is no change in the state for a free particle, $G_0\left(\mathbf{k}, t - t_1\right)$ defines the probability of transition from t_1 to t_2. Indeed, solving the time-dependent Schrödinger equation, one concludes that a free particle that was in the state $\phi_{k_1}\left(\mathbf{r}, t_1\right)$ at time t_1 will have at time t_2 the wave function $\phi_{k_1}\left(\mathbf{r}, t_1\right) \exp\left[\left(-i\varepsilon_{k_1}/\hbar\right)\left(t_2 - t_1\right)\right]$. One may expand this function in the series of wave functions ϕ_k

$$\phi_{k_1}\left(\mathbf{r}, t_1\right) \exp\left[\left(-i\varepsilon_{k_1}/\hbar\right)\left(t_2 - t_1\right)\right] = \sum_k a_k \phi_k\left(\mathbf{r}, t_2\right), \qquad (4.17)$$

with coefficients

$$a_{k_2} = \int d\mathbf{r} \phi_{k_2}^*\left(\mathbf{r}, t_2\right) \phi_{k_1}\left(\mathbf{r}, t_1\right) \exp\left[\left(-i\varepsilon_{k_1}/\hbar\right)\left(t_2 - t_1\right)\right] \qquad (4.18)$$

$$= \delta_{k_1, k_2} \exp\left[\left(-i\varepsilon_{k_1}/\hbar\right)\left(t_2 - t_1\right)\right]$$

This gives the amplitude of the transition probability, which coincides (up to the factor $i/\hbar$) with the Green function (4.14) of a free particle. The importance of the Green function lies in its time Fourier transform $G_0(k, \omega)$, whose denominator $\omega - \varepsilon_k/\hbar$ determines the energy of the particle. In the general case, the denominator of the Green function $G(k, \omega)$ gives the energies of an ideal gas of elementary excitations.

4.3   Green function in second quantization

As was the case for the S-matrix, one can write the Green function in the interaction representation. The operators $a_k\left(t\right)$ and $a_k^+\left(t\right)$ in the interaction representation are expressed in the following manner through the Schrödinger operators a_k and a_k^+,

$$a_k\left(t\right) = \exp\left(iH_0 t/\hbar\right) a_k \exp\left(-iH_0 t/\hbar\right), \qquad (4.19)$$

$$a_k^+\left(t\right) = \exp\left(-iH_0 t/\hbar\right) a_k^+ \exp\left(iH_0 t/\hbar\right)$$

Using the commutation relations $a_k a_l^+ + a_l^+ a_k = \delta_{k,\, l}$, one obtains the commutation relations for the operators $a_k\left(t\right)$ and $a_k^+\left(t\right)$,

$$a_k\left(t_1\right) a_l^+\left(t_2\right) + a_l^+\left(t_2\right) a_k\left(t_1\right) \qquad (4.20)$$

$$= \delta_{k,\, l} \exp\left[\left(-i\varepsilon_{k_1}/\hbar\right)\left(t_2 - t_1\right)\right] = i\hbar \delta_{k,\, l} G_0\left(k, t_2 - t_1\right),$$

which are expressed in terms of the Green function of a free particle.

To express the Green function in the operators $a_k\left(t\right)$ and $a_k^+\left(t\right)$, one introduces the Heisenberg operators $\hat{a}_k\left(t\right)$ and $\hat{a}_k^+\left(t\right)$,

$$\hat{a}_k\left(t\right) = \exp\left(iHt/\hbar\right) a_k \exp\left(-iHt/\hbar\right), \qquad (4.21)$$

$$\hat{a}_k^+\left(t\right) = \exp\left(-iHt/\hbar\right) a_k^+ \exp\left(iHt/\hbar\right)$$

The Green function can be constructed as

$$G\left(k_1, k_2, t - t_1\right) = \left\langle 0 \left| \exp\left(\frac{iH}{\hbar} t_2\right) \hat{a}_{k_2}\left(t_2\right) \right. \right. \tag{4.22}$$

$$\times \exp\left[-\frac{iH}{\hbar}\left(t_2 - t_1\right)\right] \hat{a}_{k_1}^+\left(t_1\right) \exp\left(-\frac{iH}{\hbar} t_1\right) \left| 0 \right\rangle$$

Starting from the vacuum state $|0\rangle$, the operator $\exp\left(-iHt_1/\hbar\right)$ transfers the system to time t_1, where the operator $\hat{a}_{k_1}^+\left(t_1\right)$ describes the appearance of a particle in state k_1. Then, the operator $\exp\left[-iH\left(t_2 - t_1\right)/\hbar\right]$ transfers the system to time t_2, where the particle appears in state k_2, which is described by the operator complex-conjugate to $\hat{a}_{k_2}^+\left(t_2\right)\exp\left(-iHt_2/\hbar\right)|0\rangle$. It only remains to replace the Heisenberg operators $\hat{a}_k\left(t\right)$ and $\hat{a}_k^+\left(t\right)$ in Eq. (4.22) by the operators $a_k\left(t\right)$ and $a_k^+\left(t\right)$ in the interaction representation. These two groups of operators are related by Eqs. (4.19) and (4.21),

$$\hat{a}_k\left(t\right) = u^*\left(t\right) a_k\left(t\right) u\left(t\right); \quad \hat{a}_k^+\left(t\right) = u\left(t\right) a_k^+\left(t\right) u^*\left(t\right) \tag{4.23}$$

where $u\left(t\right) = \exp\left[\left(it/\hbar\right)\left(H_0 - H\right)\right] = \exp\left(-itH_I/\hbar\right)$ is the solution of Eq. (4.1), which, according to (4.6), must satisfy

$$u\left(t_2\right) = S\left(t_2 - t_1\right) u\left(t_1\right) \tag{4.24}$$

Inserting (4.23) and (4.24) into (4.22) yields

$$G\left(k_1, k_2, t - t_1\right) = \left\langle 0 \left| a_{k_2}\left(t_2\right) S\left(t_2, t_1\right) a_{k_1}^+\left(t_1\right) \right| 0 \right\rangle \tag{4.25}$$

This leads to the question of how to calculate the S-matrix and Green function, which, according to Eqs. (4.8) and (4.25), are the chronological products of many operators in the interaction representation.

4.4 Wick theorem

In addition to the chronological operator $\hat{P}$ introduced above, it is convenient to define the normal operator $\hat{N}$ which arranges all the operators in such a way that $a\left(t\right)$ operators always appear after $a^+\left(t\right)$ operators. The importance of operator $\hat{N}$ lies in the fact that its action on the vacuum state vanishes since

$$a\left(t\right)|0\rangle = 0 \tag{4.26}$$

One defines the pairing of two operators $\overbrace{a_{k_2}\left(t_2\right) a_{k_1}^+}\left(t_1\right)$ as

$$\overbrace{a_{k_2}\left(t_2\right) a_{k_1}^+}\left(t_1\right) = \hat{P}\left[a_{k_2}\left(t_2\right) a_{k_1}^+\left(t_1\right)\right] - \hat{N}\left[a_{k_2}\left(t_2\right) a_{k_1}^+\left(t_1\right)\right]$$

$$= i\hbar\delta_{k_1 k_2} G_0\left(k_1, t_2 - t_1\right) \tag{4.27}$$

The latter equality follows from the definitions of operators $\hat{P}$ and $\hat{N}$ and (4.20). It is easy to verify that for all other pairs of operators $a_k(t)$ and $a_k^+(t)$, their pairing vanishes. The full proof of the generalization of Eq. (4.27) to many operators (Wick theorem) is carried out by mathematical induction [22]. The result of this calculation for a series of operators $a_k(t)$ and $a_k^+(t)$ yields

$$\hat{P}\left(a^+a^+a, ...aa^+a + ...\right) = \left(a^+\overbrace{a^+a}, ...aa + ...\right) \qquad (4.28)$$

$$+ \left(a^+\overbrace{a^+a}, ...\overbrace{aa^+aa} + ...\right) + ... \left(\overbrace{a^+aa^+a}, ...\overbrace{aa^+} + ...\right)$$

where the right-hand side contains terms with one, two ... pairings. According to Eq. (4.26), the final terms, which contain only pairing operators, will be the only ones which remain in the Green function after averaging over the vacuum state. For the simple case of four operators, Eqs. (4.28) and (4.27) yield

$$\langle 0 | P a_k^+(t_1) a_l^+(t_2) a_m(t_3) a_n(t_4) | 0 \rangle = \qquad (4.29)$$

$$\delta_{k,m}\delta_{l,n}\left[i\hbar G_0(k, t_3 - t_1)\right]\left[i\hbar G_0(l, t_4 - t_2)\right] +$$

$$\delta_{k,n}\delta_{l,m}\left[i\hbar G_0(k, t_4 - t_1)\right]\left[i\hbar G_0(l, t_3 - t_2)\right]$$

4.5   Feynman diagrams

Since the series for both the S-matrix and the Green function contain terms with products of many operators $a_k(t)$ and $a_k^+(t)$, their calculation is quite cumbersome. It significantly simplifies the calculations if one replaces the complex analytic expressions by diagrams. The utility of diagrams can be seen from the simple example of summing the series $1/2 + 1/4 + ...$. The summation of this geometric series, which equals 1, can be replaced by the simple geometric construction shown in Fig. 4.1, where the addition of half the figure, plus another half, and plus another half, etc., clearly leads to the correct result.

In our problem, one uses the coordinate-time or wave vector-time approach. The latter approach involves displaying the destruction operator $a_k(t)$ and the creation operator $a_k^+(t)$ of electrons in Fig. 4.2 by a vertical line directed upward. For positively charged particles, described by the operators $b_k(t) = a_k^+(t)$ and $b_k^+(t) = a_k(t)$ (positrons or holes in solids), the lines are directed downward. The upward directed line, which extends from time t_1 to t_2, represents the Green function $G_0(k, t_2 - t_1)$ of a free particle. We consider two types of interactions $H_I(t)$ entering

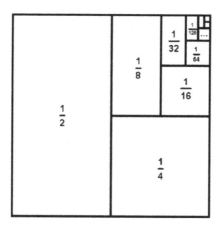

Fig. 4.1 Graphic representation for the summation of the series $1/2 + 1/4 + 1/8 + ...$

the series for $S(t_1, t_2)$ and $G(k, t_2 - t_1)$: a particle in an external field $H_I = \sum_{k,l} V_{k,l} a_k^+(t) a_l(t)$ and the interactions between electrons (or holes) $H_I = \sum_{k,l,m,n} V_{k,l,m,n} a_k^+(t) a_l^+(t) a_m(t) a_n(t)$. The first type of interaction is displayed graphically by the circle, and the second type by the wavy line. Typical diagrams are shown in Figs. 4.2 and 4.3.

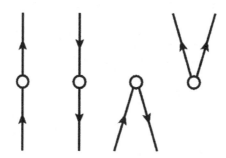

Fig. 4.2 Typical diagrams for an electron subject to an external field.

Relating each term in the series expansions of (4.8) and (4.25) to the appropriate diagram leads to two possibilities: the exact solution by summing all the diagrams or an approximate solution obtained by choosing the most important diagrams in each order of perturbation theory. We shall

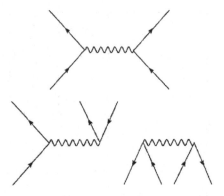

Fig. 4.3   Typical diagrams for a system of interacting electrons.

bring examples of both these approaches.

4.5.1   *Electrons in an external field*

Typical graphs of electrons (or holes) in an external field are shown in Fig. 4.4. Consider the case in which the matrix elements $V_{k,l}$ are large compared with all other matrix elements $V_{p.q}$ with $p, q \neq k, l$.

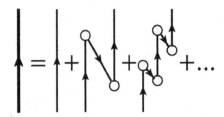

Fig. 4.4   Green function of an electron in an external field with interaction $V_{k,l}$ between states k and l.

The perturbation series for the Green function $G(k, t_2 - t_1)$, depicted in Fig. 4.4 by the thick line, equals the Green function $G_0(k, t_2 - t_1)$ of a particle not subject to an external field in the interval (t_1, t_2) plus those with two, four ... interactions. The second term in Fig. 4.4 displays the process wherein the particle with wave vector k disappears at time t' leaving a hole in state l, which is annihilated at time t'' by the incident electron. As shown in Fig. 4.4, the diagrammatic series represents a geometric progression which can be summed. The denominator of the time Fourier component $G(k, \omega)$ of the sum represents the energy of the quasiparticles of the ideal gas, which describes the excitations of the system. With our graph-formula dictionary, one can write the zero of the denominator of the formula in Fig. 4.4,

$$G_0^{-1}(k, \omega) - |V_{k,l}|^2 G_0(l, \omega) = 0 \tag{4.30}$$

or, after using (4.14) one obtains

$$(\hbar\omega - \varepsilon_k) - \frac{|V_{k,l}|^2}{\hbar\omega - \varepsilon_l} = 0 \tag{4.31}$$

Equation (4.31) closely resembles the quantum mechanical problem of two energy levels, ε_k and ε_l, with interaction $V_{k,l}$ connecting them, or the splitting of the degenerate level $\varepsilon_k = \varepsilon_l$.

4.5.2 Hartree and Hartree-Fock approximations

The diagrams for interacting electrons look quite different from the diagrams describing a particle moving in an external field, where, due to interactions, particles are able to change their momentum or annihilate leaving a hole, As shown in Fig. 4.3, the interaction between electrons (or holes) is due to the exchange of photons, shown by the wavy lines, so that at each point on the diagram, two electron (or hole) lines are joined by a wavy photon line.

The Hartree approximation is an approximate way to describe a system of interacting electrons, where each electron experiences the field created by all other electrons. Thus, the Schrödinger equation takes the following form

$$-\frac{\hbar^2}{2m}\nabla^2\Psi_k(\mathbf{r}) + \sum_l V_{k,l}\Psi_l(\mathbf{r}) = E_k\Psi_k(\mathbf{r}) \tag{4.32}$$

where the effective potential $V_{k,l}$ is given by

$$V_{k,l} = \int V(\mathbf{r} - \mathbf{r}_1)|\Psi_l(\mathbf{r}_1)|^2 d^3\mathbf{r}_1, \tag{4.33}$$

The self-consistent field $V_{k,l}$ is created by all electrons including the k-electron. Equation (4.32) can be solved by the method of variations, starting from some system of wave functions and substituting the resulting functions back into the equations until the functions obtained in the last step are very close to the functions used in the previous step.

The Hartree approximation can be formulated in the diagrammatic approach, assuming that the interaction between electrons manifests itself in scattering without a change in the wave vectors. The appropriate diagrams are shown in Fig. 4.5. The Green function $G(k,\omega)$ for such a process is described by the geometric series shown in Fig. 4.5, where the denominator of this series has the following form,

$$G_0^{-1}(k,\omega) - \sum_l V_{k,l,k,l} = 0 \qquad (4.34)$$

or

$$\hbar\omega = \varepsilon_k + \sum_l V_{k,l,k,l} \qquad (4.35)$$

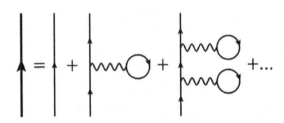

Fig. 4.5   Graphic series for the Green function in the Hartree approximation.

One can improve the Hartree approximation by adding exchange scattering to the Coulomb interaction shown in Fig. 4.5. The appropriate series will then contain combinations of two types of elementary diagrams, as shown in Fig. 4.6 (the Hartree–Fock approximations). The summation of this series leads to elementary excitations with energies

$$\hbar\omega = \varepsilon_k + \sum_l (V_{k,l,k,l} - V_{k,l,l,k}) \qquad (4.36)$$

4.5.3   *Electron gas at low and high density*

In the previous section we considered systems for which one can carry out an exact summation of the infinite series of perturbation theory, and thus obtain the exact solution of the problem. Another way of analyzing such series consists of picking out the most important terms in each order of perturbation theory and thereby obtaining an approximate sum of the entire series. Consider the special case of an electron gas situated on a background of positive charge, which provides charge stability of the system.

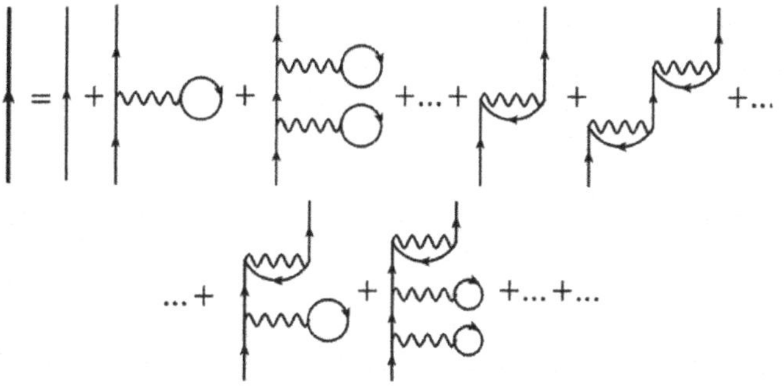

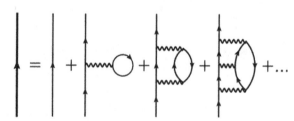

Fig. 4.6 Graphic series for the Green function in the Hartree-Fock approximation.

Such a system corresponds to electrons in a solid with fixed ions. Depending on the density n of the electron gas, there are two limiting cases of low and high densities, where the average distance between particles $n^{-1/3}$ is much larger or much smaller than the radius of the force a ($n^{-1/3} >> a$ or $n^{-1/3} << a$). The electron density is related to the value of the Fermi wave vector k_F by $n^{1/3} \sim k_F$. Hence, the electron gas at low and high density corresponds to $ak_F << 1$ and $ak_F >> 1$, respectively.

Fig. 4.7 Graph of the Green function of the electron gas at low density.

For low densities, the small parameter ak_F determines the number of holes, and, as shown in Fig. 4.7, one can identify the diagrams with the smallest number (one) of holes in each order of perturbation theory. The summation of this series can be performed [23], which gives the effective mass near the Fermi surface,

$$m_{eff} = m \left[1 + \frac{8}{15\pi^2} \left(7 \ln 2 - 1 \right) (ak_F)^2 \right] \tag{4.37}$$

We now turn to the electron gas at high density The part of the Hamiltonian H_I which describes the interaction between electrons,

$H_I = \sum V_{k,l,m,n}\, a_k^+\,(t)\, a_l^+\,(t)\, a_m\,(t)\, a_n\,(t)$ with $k+l = m+n$, can be rewritten as

$$H_I = \sum_{m,n,q} V_q a_{m-q}^+\,(t)\, a_{n+q}^+\,(t)\, a_m\,(t)\, a_n\,(t) \tag{4.38}$$

For the Coulomb interaction, $V_q \sim 1/q^2$. The term with $q = 0$ is cancelled by the positive charge background. The main contribution to the sum in Eq. (4.38) comes from small q. In Fig. 4.8, we compare the contribution of two phase diagrams of second order. As one can see from this figure, the left-hand diagram contains two photon lines with wave vector q, whereas the right-hand diagram has only one. This makes the left-hand diagram (and all similar diagrams in each order of perturbation theory) the most important, leading to the series shown in Fig. 4.9. The summation of all these diagrams results in replacing the Coulomb interaction by the screened (Yukawa) interaction [23].

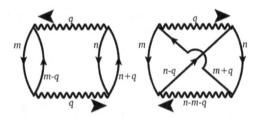

Fig. 4.8   Two graphs of second order for electron gas at high density.

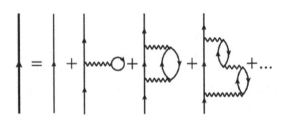

Fig. 4.9   Graph of the Green function for electron gas at high density.

4.6   Problems

Problem 4.1.

a) Which of the following diagrams shown in Fig. 4.10 appears in the Hartree-Fock approximation?

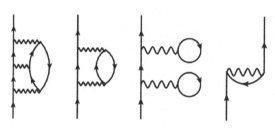

Fig. 4.10   Different Feynman diagrams.

b) For pairs of fermions interacting by a Yukawa potential, $V(r) = (V_0/r)\exp(-\alpha r)$, write the Hamiltonian in second quantization and find the energy of the quasiparticles.

c) Sum the geometric series of the Hartree diagrams and calculate the energy of the quasiparticles for the Yukawa potential.

Solution.

a) Diagrams 3 and 4.

b) The Hamiltonian in the second-quantization representation is

$$H = \sum_k E_k a_k^+ a_k + \sum_{k,l,m,n} V_{k,l,m,n} a_k^+ a_l^+ a_m a_n \qquad (4.39)$$

For plane-wave basis functions, the matrix element $V_{k,l,m,n}$ has the following form,

$$V_{k,l,m,n} = \int \exp\left[-i\mathbf{r}\cdot(\mathbf{k}+\mathbf{l}-\mathbf{m}-\mathbf{n})\right] d\mathbf{r} \qquad (4.40)$$

$$\times V_0 \int \frac{\exp(-R\alpha)}{R} \exp(-i\mathbf{q}\mathbf{R})\, d\mathbf{R}$$

$$= \delta_{k+l,m+n} V_0 \int \frac{\exp(-R\alpha)}{R} \exp(-i\mathbf{q}\mathbf{R})\, d\mathbf{R}$$

$$= \delta_{k+l,m+n} \frac{4\pi V_0}{(\mathbf{k}-\mathbf{m})^2 + \alpha^2}$$

where the integration in (4.40) has been performed by a change of variables, $\mathbf{r} = \mathbf{r}_1 - \mathbf{r}_2$ and $\mathbf{q} = \mathbf{k} - \mathbf{m}$.

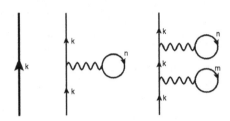

Fig. 4.11   First three Hartree diagrams.

c) The sum of the first three Hartree diagrams shown in Fig. 4.11 diagrams equals

$$G\left(k, t_2 - t_1\right) = G_0(k, t_2 - t_1) \tag{4.41}$$

$$+ \int_{t_1}^{t_2} \sum_n G_0(k, t' - t_1) G_0(k, t_2 - t') V_{k,n,k,n} dt'$$

$$+ \int_{t_1}^{t_2} \int_{t_1}^{t_2} \sum_n \sum_m G_0(k, t' - t_1) G_0(k, t'' - t')$$

$$\times G_0(k, t_2 - t'') V_{k,n,k,n} V_{k,m,k,m} dt' dt'' + \dots$$

The summation of the series can easily be performed with the help of the Feynman diagram, and its Fourier transform

$$G(k, \omega) = \frac{1}{\hbar\omega - \epsilon_k - \displaystyle\sum_{n<k_f} V_{k,n,k,n} + i\delta} \tag{4.42}$$

which gives the energy of the quasiparticles,

$$E_k = \epsilon_k + \sum_{n<k_f} V_{k,n,k,n} \tag{4.43}$$

For the Yukawa potential, the second term in (4.43) has been obtained previously, $V_{k,n,k,n} = 4\pi V_0 / \left[(k - k)^2 + \alpha^2 \right]$, and

$$\sum_{n<k_f} V_{k,n,k,n} = \sum_{n<k_f} \frac{4\pi V_0}{\alpha^2} = \frac{4\pi V_0}{\alpha^2} N_f \tag{4.44}$$

Chapter 5

Relativistic Quantum Mechanics

To establish a link between non-relativistic and relativistic quantum mechanics, one starts with the Schrödinger equation $i\hbar\partial\Psi/\partial t = -\left(\hbar^2/2m\right)\nabla^2\Psi$. Combining this equation with its complex conjugate yields the conservation law

$$\frac{\partial\rho}{\partial t} = -\nabla J; \quad \rho = \Psi\Psi^*; \quad J = \frac{\hbar}{2im}\left(\Psi^*\nabla\Psi - \Psi\nabla\Psi^*\right) \tag{5.1}$$

where the probability density $\Psi\Psi^*$ is positive, as required.

The Schrödinger equations contains the first derivative in time and the second derivative in coordinates, which makes it unsuitable for the relativistic regime. The relativistic wave equation must contain the same order of derivatives with respect to the coordinates and to time. They could be of second order or of first order, leading to the Klein-Gordon and Dirac equations, respectively. These two equation will be considered in Sections 5.1 and 5.2. The rest of this Chapter is devoted to the new results that follow from the Dirac equation for particles with spin 1/2. These include "zitterbewegung" (Section 5.3), spin magnetic and mechanical moments (Section 5.4), and some unusual effects (Klein paradox, Lamb shift, Casimir force), which are connected with the surprising nature of the "physical vacuum".

5.1 Klein-Gordon equation

As described above, one can use the dispersion relation (in this case - relativistic) to obtain the quantum equation of motion. The relativistic dispersion equation has the form $E = \sqrt{p^2c^2 + m^2c^4}$, which, for non-relativistic energies, reduces to the rest energy mc^2 plus the kinetic energy $p^2/2m$,

$$E = mc^2\sqrt{1 + \frac{p^2}{m^2c^2}} \approx mc^2 + \frac{p^2}{2m} \tag{5.2}$$

Replacing E and p in the equation $E = \sqrt{p^2 c^2 + m^2 c^4}$ by the appropriate operators, leads to

$$-\hbar^2 \frac{\partial^2 \Psi}{\partial t^2} = -c^2 \hbar^2 \nabla^2 \Psi + m^2 c^4 \Psi \qquad (5.3)$$

As in the non-relativistic case, Eq. (5.3) and the complex conjugate equation yield the conservation law

$$\frac{\partial \rho}{\partial t} = -\nabla J, \qquad \rho = \frac{i\hbar}{2mc^2} \left(\Psi^* \frac{\partial \Psi}{\partial t} - \Psi \frac{\partial \Psi^*}{\partial t} \right), \qquad (5.4)$$

$$J = \frac{\hbar}{2im} \left(\Psi^* \nabla \Psi - \Psi \nabla \Psi^* \right)$$

The expression for the probability current J agrees with that obtained in Eq. (5.1) for the non-relativistic case, but the expression for the probability density ρ is different. Whereas the probability density in the non-relativistic case is always positive, the relativistic probability density (5.4) may be negative, for certain values of the wave functions and their derivatives. Another drawback of the Klein-Gordon equation (5.3) is the absence of spin, which means that this equation can only describe a particle with zero spin.

5.2   Dirac equation

The lack of spin in the Schrödinger equation is an essential defect. Pauli suggested overcoming this difficulty for the electron in a magnetic field by adding (in addition to usual replacement $\mathbf{p}$ by $\mathbf{p} - e\mathbf{A}/c$) the 2×2 matrix σ, which transforms the Schrödinger equation into the following form

$$i\hbar \frac{\partial \Psi}{\partial t} = \frac{1}{2m} \left[\sigma \left(\mathbf{p} - \frac{e\mathbf{A}}{c} \right) \right]^2 \Psi \qquad (5.5)$$

where

$$\sigma_x = \begin{pmatrix} 0 & 1 \\ 1 & 0 \end{pmatrix}; \qquad \sigma_y = \begin{pmatrix} 0 & -i \\ i & 0 \end{pmatrix}; \qquad \sigma_z = \begin{pmatrix} 1 & 0 \\ 0 & -1 \end{pmatrix} \qquad (5.6)$$

It is readily shown that

$$\sigma_i^2 = 1; \qquad \sigma_i \sigma_j + \sigma_j \sigma_i = 0; \qquad \sigma_i \sigma_j - \sigma_j \sigma_i = 2i\sigma_k \qquad (5.7)$$

where i, j, k means the cyclic order x, y, z.

Such an addition to the Schrödinger equation indeed leads to the appearance of spin. The latter comes from the equality $p^2 = (\sigma_x p_x + \sigma_y p_y + \sigma_z p_z)^2$.

However, in the relativistic case, one cannot express $\sqrt{c^2p^2 + m^2c^4}$ as a linear combination of the 2×2 matrices, and one has to use 4×4 matrices. This was done by Dirac, who suggested the relativistic Hamiltonian of the form

$$H = c\alpha p + \beta mc^2 \qquad (5.8)$$

where

$$\alpha_x = \begin{pmatrix} 0 & \sigma_x \\ \sigma_x & 0 \end{pmatrix}; \qquad \alpha_y = \begin{pmatrix} 0 & \sigma_y \\ \sigma_y & 0 \end{pmatrix}, \qquad (5.9)$$

$$\alpha_z = \begin{pmatrix} 0 & \sigma_z \\ \sigma_z & 0 \end{pmatrix}; \qquad \beta = \begin{pmatrix} I & 0 \\ 0 & -I \end{pmatrix},$$

and I is the 2×2 unit matrix. One can easily obtain the following relations

$$\alpha_i^2 = 1; \qquad \alpha_i\alpha_j + \alpha_j\alpha_i = 0; \qquad (5.10)$$

$$\alpha_i\alpha_j - \alpha_j\alpha_i = 2i\alpha_k; \qquad \alpha_i\beta + \beta\alpha_i = 0$$

For a free particle, $H\Psi = E\Psi$, the energy E and momentum p are conserved, and the wave equation has the following form

$$\left(E - c\alpha p - \beta mc^2\right)\Psi = 0 \qquad (5.11)$$

Multiplying this equation by $E + c\alpha p + \beta mc^2$ and using the relation (5.10) leads to the correct dispersion relation,

$$E = \pm\sqrt{p^2c^2 + m^2c^4} \qquad (5.12)$$

As we shall see, there is no need, as is usually done, to reject the negative energies in (5.12).

5.3   Dynamic solution of the Dirac equation

The wave equation associated with the Hamiltonian (5.8) has the following form

$$i\hbar\frac{\partial\Psi}{dt} = \left(c\alpha p + \beta mc^2\right)\Psi \qquad (5.13)$$

Combining Eq. (5.13) with its complex conjugate, one obtains the conservation law

$$\frac{\partial\rho}{\partial t} = -\nabla J; \qquad \rho = \Psi\Psi^*; \qquad J = \Psi^*c\alpha\Psi \qquad (5.14)$$

In contrast to the Klein-Gordon equation, the Dirac equation leads to a positive density ρ. One can appreciate the last equation by passing to dynamic variables. The operator for the dynamic behavior of a physical quantity a is defined by the Poisson bracket of the operator $\hat{a}$ and Hamiltonian H,

$$\frac{d\hat{a}}{dt} = \frac{i}{\hbar} (H\hat{a} - \hat{a}H) \qquad (5.15)$$

For the coordinate x and the matrix α_x, one obtains from (5.10)

$$\frac{dx}{dt} = c\alpha_x; \qquad \frac{d\alpha_x}{dt} = \frac{2i}{\hbar} (H\alpha_x - cp_x) \qquad (5.16)$$

Differentiating the second equation in (5.16) with respect to t and solving for $d\alpha_x/dt$ yields

$$\frac{d\alpha_x}{dt} = \left(\frac{d\alpha_x}{dt} \right)_{t=0} \exp\left(i\frac{2Ht}{\hbar} \right) \qquad (5.17)$$

Inserting (5.17) into (5.16) and solving for α_x gives

$$\alpha_x = \frac{cp_x}{H} + \frac{\hbar}{2iH} \left(\frac{d\alpha_x}{dt} \right)_{t=0} \exp\left(i\frac{2H}{\hbar}t \right) \qquad (5.18)$$

Finally, the first equation in (5.16), together with Eq. (5.18), leads to the following law of motion

$$x = \frac{c^2 p_x}{H} t - \frac{\hbar^2 c}{4H^2} \left(\frac{d\alpha_x}{dt} \right)_{t=0} \exp\left(i\frac{2H}{\hbar}t \right) \qquad (5.19)$$

The first term in Eq. (5.19) shows that the velocity v equals $c^2 p_x/H$, which is quite different from the usual $v = p/m$. The second term is even more surprising. It shows that in addition to uniform motion along the x axis, the particle performs transverse oscillations with a very small amplitude of 10^{-15} cm but a very high frequency $2H/\hbar \approx 10^{21}$ sec^{-1}. These oscillations are called "zitterbewegung", which in translation from German (the physics language of 80 years ago) means "trembling motion". Since these amplitudes are too small and the frequencies too high for experimental observation, "zitterbewegung" remained a curiosity for a long time. However, recent experiments [24] have been performed with a single non-relativistic ion trapped in an electromagnetic cage by a laser field. It turns out that the Schrödinger equation that describes this ion is identical to the Dirac equation for the relativistic electron. The zitterbewegung of an ion has the much larger amplitude of a few nanometers. Researchers [24] used a calcium ion, studied its motion transferred to the internal state of the ion and measured the emitted fluorescent light. Changing the effective mass of the ion and keeping its momentum constant, one can go from the non-relativistic regime (large effective mass) to the highly relativistic regime (small effective mass). Figure 5.1 shows the average position of the ion for different values of mass. The lightest line corresponds to zero mass, and its trajectory is the straight line. For higher mass, the trajectory is no longer a straight line, but exhibits zitterbewegung.

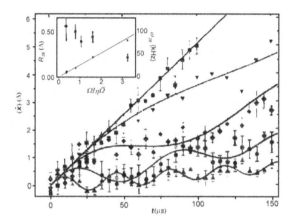

Fig. 5.1 Expectation values for particles with varying masses. The linear curve (squares) represents a massless particle moving at the speed of light. The other curves represent particles of increasing mass with different Compton wavelengths. The solid curves represent numerical simulations. The figure shows zitterbewegung at the crossover from the relativistic to the non-relativistic regimes. The inset displays the fitted zitterbewegung amplitude versus the parameter proportional to the mass. The error bars correspond to one standard deviation. [Reprinted from R. Gerritsma *et al.*, Nature **463**, 68 (2010), Copyright (2010) Nature Publishing Group]

5.4 Electron spin

The mechanical and magnetic moment of an electron (spin) appear naturally in the Dirac equation. One starts from the mechanical moment of a particle. In the Schrödinger equation, the angular momentum $\mathbf{L} = \mathbf{r} \times \mathbf{p}$ is a conserved quantity. In order to examine the situation corresponding to the relativistic Dirac equation (5.13), one writes Eq. (5.15) with $a = L_x$,

$$\frac{dL_x}{dt} = \frac{i}{\hbar} \{H, L_x\} = \frac{ic}{\hbar} \left[\alpha_y \{p_y y\} p_z - \alpha_z \{p_z z\} p_y \right] \qquad (5.20)$$

$$= c \left(\alpha_y p_z - \alpha_z p_y \right) \neq 0,$$

where we used the commutation rule for the momentum and coordinate, $\{p_y, y\} = \{p_z, z\} = -i\hbar$. As one sees from Eq. (5.20), the angular momentum $\mathbf{L}$ is not conserved in relativistic quantum mechanics.

Let us define the 4×4 matrices $\overline{\sigma}_i$,

$$\overline{\sigma}_x = \begin{pmatrix} \sigma_x & 0 \\ 0 & \sigma_x \end{pmatrix}, \quad \overline{\sigma}_y = \begin{pmatrix} \sigma_y & 0 \\ 0 & \sigma_y \end{pmatrix}; \quad \overline{\sigma}_z = \begin{pmatrix} \sigma_z & 0 \\ 0 & \sigma_z \end{pmatrix} \qquad (5.21)$$

These matrices satisfy the following commutation relations with the matrices α_i and β, defined in Eqs. (5.9) and (5.10)

$$\overline{\sigma}_x \alpha_y + \alpha_y \overline{\sigma}_x = 0; \quad \overline{\sigma}_x \alpha_z + \alpha_z \overline{\sigma}_x = 0; \quad \overline{\sigma}_x \beta + \beta \overline{\sigma}_x = 0;$$

$$\{\alpha_y \overline{\sigma}_x\} = \alpha_y \overline{\sigma}_x - \overline{\sigma}_x \alpha_y = -2i\sigma_z; \qquad (5.22)$$

$$\{\alpha_z \overline{\sigma}_x\} = \alpha_z \overline{\sigma}_x - \overline{\sigma}_x \alpha_z = 2i\sigma_y; \quad \alpha_x \alpha_y = i\overline{\sigma}_z$$

Inserting $a = \overline{\sigma}_x$ into Eq. (5.15) and using Eq. (5.22), one obtains

$$\frac{d\overline{\sigma}_x}{dt} = \frac{i}{\hbar} \{H, \overline{\sigma}_x\} = \frac{ic}{\hbar} \left[p_y \{\alpha_y \overline{\sigma}_x\} + p_z \{\alpha_z \overline{\sigma}_x\} \right] \qquad (5.23)$$

$$= \frac{2c}{\hbar} (\alpha_z p_y - \alpha_y p_z)$$

Combining Eqs. (5.20) and (5.23) leads to

$$\frac{d}{dt} \left(\mathbf{L} + \frac{\hbar}{2}\overline{\boldsymbol{\sigma}} \right) = 0 \qquad (5.24)$$

Thus, in relativistic quantum mechanics, the conserved quantity is not the angular momentum $\mathbf{L}$ but rather the sum of the angular momentum $\mathbf{L}$ and the intrinsic momentum $\hbar\overline{\boldsymbol{\sigma}}/2$ of the particle.

In addition to the extra mechanical moment $\hbar\overline{\boldsymbol{\sigma}}/2$, a quantum particle described by the Dirac equation also possesses an intrinsic magnetic moment (spin). To describe the behavior of a particle in a magnetic field, one has to replace $\mathbf{p}$ in Eq. (5.8) by $\mathbf{p} - e\mathbf{A}/c$, where $\mathbf{A}$ is the vector potential of an external magnetic field, $\mathbf{B} = \nabla \times \mathbf{A}$. Multiplying the modified equation (5.8) $E - \alpha(pc - eA) - \beta mc^2 = 0$ by $E + \alpha(pc - eA) + \beta mc^2$ and using the commutation relations (5.22) leads to

$$(E - m^2 c)(E + mc^2) = \alpha(\mathbf{pc} - e\mathbf{A})\alpha(\mathbf{pc} - e\mathbf{A}) \qquad (5.25)$$

The right-hand side of this equation has the form

$$(\alpha\mathbf{Q})(\alpha\mathbf{R}) = (\alpha_x Q_x + \alpha_y Q_y + \alpha_z Q_z)$$
$$\times (\alpha_x R_x + \alpha_y R_y + \alpha_z R_z), \qquad (5.26)$$

which can be evaluated using the commutation relations (5.10) and (5.22),

$$(\alpha\mathbf{Q})(\alpha\mathbf{R}) = \mathbf{QR} + i\overline{\sigma}(\mathbf{Q} \times \mathbf{R}) \qquad (5.27)$$

Choosing $\mathbf{Q} = \mathbf{R} = \mathbf{pc} - e\mathbf{A}$ and inserting the results into (5.25) yields

$$\left(E - m^2 c\right)\left(E + mc^2\right) = \left(c\mathbf{p} - e\mathbf{A}\right)^2 + e\hbar c \overline{\boldsymbol{\sigma}}\mathbf{B} \qquad (5.28)$$

Using Eq. (5.2) with $p^2/2m \ll mc^2$ (non-relativistic quantum mechanics) leads to

$$\frac{p^2}{2m} = \frac{(\mathbf{p} - e\mathbf{A}/c)^2}{2m} + \frac{e\hbar}{2mc}\overline{\boldsymbol{\sigma}}\mathbf{B} \qquad (5.29)$$

Therefore, starting from the Dirac equation for a relativistic particle in a magnetic field, one obtains the non-relativistic Schrödinger equation for low momentum, but with an additional term that describes the interaction of the internal magnetic moment $e\hbar\overline{\boldsymbol{\sigma}}/2mc$ of the particle with an external magnetic field $\mathbf{B}$. These results for a magnetic moment $e\hbar\overline{\boldsymbol{\sigma}}/2mc$ and a mechanical moment $\hbar\overline{\boldsymbol{\sigma}}/2$ show that the Dirac equation explains the observed intrinsic moment of a particle having the gyromagnetic ratio e/mc, which equals twice the usual value.

5.5 Dirac equation for a free particle

For a free particle, both the energy E and momentum $\mathbf{p}$ are conserved, and the wave function has the form $\Psi_i = u_i \exp\left[i\left(\mathbf{kr} - \omega t\right)\right]$. Indeed, the operators for the energy $i\hbar\partial/\partial t$ and momentum $-i\hbar\nabla$ acting on this function give the correct values $\hbar\omega$ and $\hbar\mathbf{k}$. In addition, Eq. (5.11) contains 4×4 matrices α and β, and, therefore, represents four equations for the u_i, $i = 1...4$. Using the definition (5.6)-(5.9) of the matrices α and β, one can rewrite Eqs. (5.11) in the following form,

$$E\begin{pmatrix} u_1 \\ u_2 \\ u_3 \\ u_4 \end{pmatrix} - cp_x\begin{pmatrix} 0\,0\,0\,1 \\ 0\,0\,1\,0 \\ 0\,1\,0\,0 \\ 1\,0\,0\,0 \end{pmatrix}\begin{pmatrix} u_1 \\ u_2 \\ u_3 \\ u_4 \end{pmatrix}$$

$$- cp_y\begin{pmatrix} 0\;\;0\;\;0\;-i \\ 0\;\;0\;\;i\;\;0 \\ 0\,-i\;0\;\;0 \\ i\;\;0\;\;0\;\;0 \end{pmatrix}\begin{pmatrix} u_1 \\ u_2 \\ u_3 \\ u_4 \end{pmatrix} - cp_z\begin{pmatrix} 0\;\;0\;\;1\;\;0 \\ 0\;\;0\;\;0\,-1 \\ 1\;\;0\;\;0\;\;0 \\ 0\,-1\;0\;\;0 \end{pmatrix}\begin{pmatrix} u_1 \\ u_2 \\ u_3 \\ u_4 \end{pmatrix}$$

$$+mc^2\begin{pmatrix} 1\;\;0\;\;0\;\;0 \\ 0\;\;1\;\;0\;\;0 \\ 0\;\;0\,-1\;\;0 \\ 0\;\;0\;\;0\,-1 \end{pmatrix}\begin{pmatrix} u_1 \\ u_2 \\ u_3 \\ u_4 \end{pmatrix} = 0 \qquad (5.30)$$

which represents the following four equations,

$$\left(E + mc^2\right) u_1 - c\left(p_x - ip_y\right) u_4 - cp_z u_3 = 0, \qquad (5.31)$$
$$\left(E + mc^2\right) u_2 - c\left(p_x + ip_y\right) u_3 + cp_z u_4 = 0,$$
$$\left(E - mc^2\right) u_3 - c\left(p_x - ip_y\right) u_2 - cp_z u_1 = 0,$$
$$\left(E - mc^2\right) u_4 - c\left(p_x + ip_y\right) u_1 + cp_z u_2 = 0$$

The eigenvalues of Eqs. (5.31) are given by the zeros of the determinant of these equations, which after simple but tedious calculation yields two solutions,

$$E_+ = +\sqrt{c^2 p^2 + m^2 c^4}; \qquad E_- = -\sqrt{c^2 p^2 + m^2 c^4} \qquad (5.32)$$

Inserting E_+ into (5.31) leads to the following two solutions,

$$u_1 = 1, \quad u_2 = 0, \quad u_3 = \frac{cp_z}{E_+ + mc^2}, \quad u_4 = \frac{c\left(p_x + ip_y\right)}{E_+ + mc^2}, \qquad (5.33)$$

and

$$u_1 = 0, \quad u_2 = 1, \quad u_3 = \frac{c\left(p_x - ip_y\right)}{E_+ + mc^2}, \quad u_4 = \frac{-cp_z}{E_+ + mc^2} \qquad (5.34)$$

Similarly, one obtains for E_-,

$$u_3 = 1, \quad u_4 = 0, \quad u_1 = \frac{cp_z}{E_- - mc^2}, \quad u_2 = \frac{c\left(p_x + ip_y\right)}{E_- - mc^2}, \qquad (5.35)$$

and

$$u_3 = 0, \quad u_4 = 1; \quad u_1 = \frac{c\left(p_x - ip_y\right)}{E_- - mc^2}; \quad u_2 = \frac{-cp_z}{E_- - mc^2} \qquad (5.36)$$

In the non-relativistic limit, $v \ll c$, the two last solutions in the equations (5.33)-(5.36) are of the order v/c and may be neglected. Thus, for the eigenvalues E_+ and E_-, there are two solutions for the functions u_i. According to Eq. (5.21), these two solutions, zero and unity, correspond to two values of the spin variable, $\pm\hbar\bar{\sigma}_z/2$, as should be the case.

5.6   Motion in a central field

As for the non-relativistic Schrödinger equation, one may write the Dirac equation in spherical coordinates, introducing the radial components α_r and p_r

$$\alpha_r = \frac{\boldsymbol{\alpha r}}{r}; \quad p_r = \frac{\mathbf{pr}}{r} = \frac{\mathbf{rp} - i\hbar}{r} = -i\hbar\left(\frac{\partial}{\partial r} + \frac{1}{r}\right) \qquad (5.37)$$

Choosing $\mathbf{Q} = \mathbf{r}$ and $\mathbf{R} = \mathbf{p}$ in Eq. (5.27) gives

$$(\boldsymbol{\alpha}\mathbf{r})(\boldsymbol{\alpha}\mathbf{p}) = \mathbf{rp} + i\boldsymbol{\sigma}\mathbf{L} = \mathbf{pr} + i(\boldsymbol{\sigma}\mathbf{L} + \hbar) \tag{5.38}$$

It is convenient to introduce into the Dirac Hamiltonian the conserved variable k, defined by

$$\hbar k = \beta(\boldsymbol{\sigma}\mathbf{L} + \hbar) \tag{5.39}$$

Taking the square of (5.39) and using (5.27) to calculate $(\boldsymbol{\sigma}\mathbf{L})^2$ yields

$$\hbar^2 k^2 = \left(\mathbf{L} + \frac{\hbar}{2}\boldsymbol{\sigma}\right)^2 + \frac{\hbar^2}{4} \tag{5.40}$$

Denoting by $\hbar^2 j(j+1)$ the eigenvalues of the square of the total momentum $\mathbf{L} + \hbar\boldsymbol{\sigma}/2$, one obtains from Eq. (5.40),

$$k^2 = j(j+1) + 1/4 = \left(j + \frac{1}{2}\right)^2, \tag{5.41}$$

where k takes the integer values $\pm 1, \pm 2....$

Using (5.37)-(5.39) to replace $\boldsymbol{\sigma}\mathbf{L} + \hbar$ in Eq. (5.38) by $\beta\hbar k$, and $\mathbf{pr}$ by rp_r, leads to

$$(\boldsymbol{\alpha}\mathbf{r})(\boldsymbol{\alpha}\mathbf{p}) = rp_r + i\hbar k\beta \tag{5.42}$$

which yields, after multiplying by $\boldsymbol{\alpha}\mathbf{r} = r\alpha_r$,

$$(\boldsymbol{\alpha}\mathbf{p}) = p_r\alpha_r + \frac{i\hbar k\alpha_r\beta}{r} \tag{5.43}$$

Inserting Eqs. (5.37) and (5.43) into (5.11) permits one to rewrite the Dirac equation for a central field $V(r)$,

$$\left[E + i\hbar c\alpha_r\left(\frac{\partial}{\partial r} + \frac{1}{r}\right) - \frac{i\hbar c k\alpha_r\beta}{r} - \beta mc^2 - V(r)\right]\Psi = 0 \tag{5.44}$$

As seen in the previous section, in the non-relativistic limit, the state of a system is defined by two components of the wave function. The 2×2 matrices α_r and β, which satisfy the conditions $\alpha_r^2 = \beta^2 = 1$ and $\alpha_r\beta = -\beta\alpha_r$, can be chosen as

$$\alpha_r = \begin{pmatrix} 0 & -i \\ i & 0 \end{pmatrix}; \qquad \beta = \begin{pmatrix} 1 & 0 \\ 0 & -1 \end{pmatrix} \tag{5.45}$$

Let us seek solutions of the two equations (5.44) in the form

$$\Psi(r) = \begin{pmatrix} F/r \\ G/r \end{pmatrix} \tag{5.46}$$

Inserting (5.45) and (5.46) into (5.44) gives

$$\left(E - mc^2 - V\right) F + \frac{\hbar c k}{r} G + \hbar c \frac{dG}{dr} = 0 \qquad (5.47)$$

$$\left(E + mc^2 - V\right) G + \frac{\hbar c k}{r} F - \hbar c \frac{dF}{dr} = 0$$

For the special case $V = -Ze^2/r$, corresponding to an electron in the outer orbit of an atom with Z protons, one can rewrite Eqs. (5.47),

$$\left(\frac{d}{d\rho} - \frac{k}{\rho}\right) F - \left(\frac{\alpha_1}{\alpha} - \frac{\gamma}{\rho}\right) G = 0 \qquad (5.48)$$

$$\left(\frac{d}{d\rho} + \frac{k}{\rho}\right) G - \left(\frac{\alpha_2}{\alpha} + \frac{\gamma}{\rho}\right) G = 0$$

where

$$\alpha_1 = \frac{mc^2 + E}{\hbar c}; \qquad \alpha_2 = \frac{mc^2 - E}{\hbar c}; \qquad (5.49)$$

$$\alpha = \sqrt{\alpha_1 \alpha_2}; \qquad \rho = \alpha r; \qquad \gamma = \frac{Ze^2}{\hbar c}$$

As in the non-relativistic regime, one seeks solutions of Eqs. (5.48) of the form

$$F(\rho) = \rho^s \sum_{n=0}^{\infty} a_n \rho^n \exp(-\rho); \qquad G(\rho) = \rho^s \sum_{n=0}^{\infty} b_n \rho^n \exp(-\rho) \qquad (5.50)$$

Inserting Eq. (5.50) into (5.48) and combining the coefficients of ρ^{s+n-1} gives

$$(s + n + k) b_n - b_{n-1} + \gamma a_n - \frac{\alpha_2}{\alpha} a_{n-1} = 0 \qquad (5.51)$$

$$(s + n - k) a_n - a_{n-1} - \gamma b_n - \frac{\alpha_1}{\alpha} b_{n-1} = 0$$

Equations (5.51) for $n = 0$ have a non-zero solution when the determinant of these equations vanishes. This gives

$$s = \sqrt{k^2 - \gamma^2} \qquad (5.52)$$

For non-zero n, multiplying these equations by α and α_2, respectively, and combining them leads to

$$b_n \left[\alpha \left(s + n + k\right) + \alpha_2 \gamma\right] = a_n \left[\alpha_2 \left(s + n - k\right) - \alpha \gamma\right] \qquad (5.53)$$

For large ρ, the behavior of the series (5.50) is determined by large $n \gg s$. Equations (5.51) and (5.53) yield

$$a_n \approx \frac{2}{n} a_{n-1}; \qquad b_n \approx \frac{2}{n} b_{n-1}, \qquad (5.54)$$

implying that both series (5.50) behave as $\exp(\rho)$. To get a finite result at large ρ, one has to terminate the series at some $m = n - 1$. Therefore, Eqs. (5.51) yield $\alpha_2 a_m = -\alpha b_m$. Inserting into (5.53) with $n = N$, gives

$$4\alpha^2 (s + N)^2 = \gamma^2 (\alpha_1 - \alpha_2)^2 \tag{5.55}$$

or, according to (5.49),

$$E = mc^2 \left[1 + \frac{\gamma^2}{(s + N)^2} \right]^{-1/2} \tag{5.56}$$

Expanding in γ^2 and noting that according to equations (5.41) and (5.52), $k = |j + 1/2|$ and $s = \sqrt{k^2 - \gamma^2}$, one obtains

$$E = mc^2 \left[1 - \frac{\gamma^2}{2N^2} - \frac{\gamma^4}{2N^4} \left(\frac{N}{|j + 1/2|} - \frac{3}{4} \right) \right] \tag{5.57}$$

The first two terms in the square brackets coincide with the result of the non-relativistic theory where the energy is defined by the principal quantum number N, while the next term, which depends on the total angular momentum quantum number j, gives relativistic corrections. Therefore, energy levels such as $2S_{j=1/2}$ and $2P_{j=3/2}$ with different quantum numbers j are split ("fine structure"). This splitting is observed in precise spectroscopic experiments.

However, the levels $2S_{j=1/2}$ and $2P_{j=1/2}$, which have the same quantum numbers N and j, remain degenerate. As we will see in Section 5.7, their splitting is connected with vacuum fluctuations of the electromagnetic field.

5.7 Nature of the physical vacuum

We have seen that the Dirac equation has solutions for both positive and negative energies. In this section we consider three non-trivial effects connected with the properties of the vacuum. It turns out that the vacuum is not "empty", but is full of electron-positron pairs, which are "polarized" in the presence of a system of charges and thereby influence its behavior. To put it another way, the energy of the electromagnetic field is the sum of the energies of elementary excitations (photons), which are represented as harmonic oscillators with zero-point energies $\sum_k \hbar \omega_k / 2$. The interaction of system of charges with these "vacuum fluctuations" creates an additional force which affects the behavior of the system.

5.7.1 *Lamb shift*

The zero-point fluctuations of the electromagnetic field are given by

$$\langle E^2 \rangle = \sum_k \frac{\hbar \omega_k}{2} = 4\pi \frac{\hbar c}{2} \int k^3 dk \qquad (5.58)$$

According to the convolution theorem for the Fourier component of the fluctuations,

$$\langle E_\omega^2 \rangle = \frac{2\pi \hbar \omega^3}{c^3} \qquad (5.59)$$

Following Newton's law, $md^2\xi/dt^2 = eE$, the fluctuations of the electromagnetic field induce fluctuations in the coordinate ξ of a free electron,

$$\langle \xi^2 \rangle = \int \langle \xi_\omega^2 \rangle \, d\omega = \frac{2\pi \hbar e^2}{m^2 c^3} \int \frac{d\omega}{\omega} \qquad (5.60)$$

The last integral diverges at both the lower and upper limits of integration, and one has to cut off the limits. It is natural to take the upper (macroscopic) boundary at $\hbar\omega = mc^2$, and the lower (microscopic) boundary at a typical energy of an atomic electron $\hbar\omega = me^4/\hbar^2 = \left(e^2/\hbar c \right)^2 mc^2$. Equation (5.60) then gives

$$\langle \xi^2 \rangle = \frac{2\pi \hbar e^2}{m^2 c^3} \ln \left(\frac{1}{137} \right)^2 \qquad (5.61)$$

where the fine structure constant $e^2/\hbar c \approx 1/137$ has been introduced.

These fluctuations induce an additional averaged potential acting on the electron,

$$\langle V(r + \xi) \rangle = V(r) + \frac{1}{2} \langle \xi^2 \rangle \nabla^2 V \langle \cos^2 \theta \rangle + \ldots \qquad (5.62)$$

$$= V(r) + \frac{1}{6} \langle \xi^2 \rangle 4\pi Z e^2 \delta(r)$$

from Poisson's equation $\nabla^2 V = 4\pi Z e^2 \delta(r)$ and $\langle \cos^2 \theta \rangle = 1/3$.

The shift in the energy level, called the Lamb shift, is caused by the shift of the potential (5.62). A perturbation theory calculation yields

$$\Delta E = \frac{2\pi Z e^2 \langle \xi^2 \rangle}{3} \int \Psi_n^*(\mathbf{r}) \delta(\mathbf{r}) \Psi_n(\mathbf{r}) \, d\mathbf{r} = \frac{2\pi Z e^2 \langle \xi^2 \rangle}{3} |\Psi_n(0)|^2 \quad (5.63)$$

It is seen that the Lamb shift occurs only for spherically symmetric S-states for which $\Psi_n(0) \neq 0$.

This small energy shift was confirmed experimentally in 1947 [25]. Microwave radiation impinged on a beam of hydrogen atoms in the $2S_{j=1/2}$

state, which causes transitions to the $2P_{j=1/2}$ state. The radiation frequency implies a small (4×10^{-6} eV) shift in energy between the $2S_{j=1/2}$ and $2P_{j=1/2}$ states. These experiments stimulated the development of quantum electrodynamics. It should be mentioned that although we have used the foregoing explanation as an illustration of the nature of the physical vacuum, there are other interpretations of this effect. Particular attention has been given to graphene which is very popular these days because the electron equation of motion is very similar to the Dirac equations. The details can be found in the article [26], whose authors include two 2010 Nobel prize winners in physics.

5.7.2 Klein paradox

The essence of the Klein paradox lies in the fact that when relativistic electrons are incident on a step potential barrier (Fig. 5.2), more electrons are reflected from the barrier than are incident upon it. This paradox can be explained in the framework of the Klein-Gordon equation. The detailed analysis based on the Dirac equation and related articles can be found in [27].

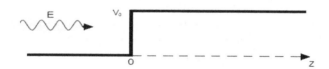

Fig. 5.2 Particles incident on the potential barrier.

In the presence of the potential step V_0, the Klein-Gordon equation (5.3) has the following one-dimensional form,

$$-\hbar^2\frac{\partial^2\Psi}{\partial t^2} = -c^2\hbar^2\frac{\partial^2\Psi}{\partial z^2} + m^2c^4\Psi; \qquad z < 0 \qquad (5.64)$$

and

$$\left(i\hbar\frac{\partial}{\partial t} - V_0\right)^2\Psi = -c^2\hbar^2\frac{\partial^2\Psi}{\partial z^2} + m^2c^4\Psi; \qquad z > 0 \qquad (5.65)$$

In the region $z < 0$, the solution has the following form, taking into account the reflected wave,

$$\Psi_1 = \exp\left(-\frac{iEt}{\hbar}\right)\left[\exp\left(\frac{ip_z z}{\hbar}\right) + B\exp\left(-\frac{ip_z z}{\hbar}\right)\right] \tag{5.66}$$

whereas in the region $z > 0$, the solution for the transmitted wave is

$$\Psi_2 = C\exp\left(-\frac{iEt}{\hbar}\right)\exp\left(\frac{iq_z z}{\hbar}\right) \tag{5.67}$$

where

$$cq_z = \pm\left[(E - V_0)^2 - m^2 c^4\right]^{1/2} = \tag{5.68}$$

$$\pm\left[(E - V_0 - mc^2)(E - V_0 + mc^2)\right]^{1/2}$$

If $E - V_0 - mc^2 < 0$ and $E - V_0 + mc^2 > 0$, q_z is imaginary. The boundary conditions $\Psi_1 = \Psi_2$ and $\partial\Psi_1/\partial t = \partial\Psi_2/\partial t$ at $z = 0$, result in

$$C = \frac{2p_z}{p_z + q_z}; \qquad B = \frac{p_z - q_z}{p_z + q_z} \tag{5.69}$$

For imaginary q_z, the reflection coefficient $|B|^2$ equals unity but for negative q_z, $|B|^2 > 1$, which is the Klein paradox.

The explanation of the Klein paradox is the following. Electrons impinging on the barrier with large energies are able to excite electron-positron pairs from the vacuum. The positrons move to the right and the excited electrons move to the left, thereby increasing the number of reflected electrons. The situation is similar to electron-hole pairs in solids with the essential difference being that after the 1929 Klein theory [28], positrons were discovered in cosmic radiation.

5.7.3   *Casimir force*

Another manifestation of the physical vacuum is the existence of an attractive force between two plates a few micrometers apart in the absence of any electromagnetic field. This phenomenon is analogous to the attraction between two parallel ships that is due to the difference in water pressure acting on the ships from outside and from between the ships. The plates attract each other, just as two objects joined by a stretched spring will move together as the energy stored in the spring decreases. In the relativistic case, the properties of the vacuum are the source of the attraction between the two plates. As the two plates move closer together, the total energy in the vacuum between the plates becomes a bit less than the energy elsewhere in the vacuum. Therefore, the plates attract each other.

For two metallic plates at a distance a apart, the transverse component of the vacuum electric field and the normal component of the vacuum magnetic field must vanish at the surface of the plates located at $z = 0$ and $z = a$. The electric field component of the standing wave between the plates has the following form,

$$\Psi_n(x, y, z, t) = \exp\left(-i\omega_n t + ik_x x + ik_y y\right) \sin\left(\frac{n\pi}{a} z\right) \tag{5.70}$$

where n is an integer. The dispersion relation of this wave is

$$\omega_n = c\sqrt{k_x^2 + k_y^2 + \frac{n^2\pi^2}{a^2}} \tag{5.71}$$

The vacuum zero-point energy is the sum over all excitations,

$$E = 2\frac{\hbar}{2} \int \frac{dk_x dk_y}{(2\pi)^2} \sum_{n=1}^{\infty} S \frac{\omega_n}{\omega^l} \tag{5.72}$$

where S is the area of the plates and an additional factor 2 arises from the two polarizations of the wave. An index l has been introduced in (5.72), which will be taken to be $l = 0$ in the end of the calculations.

Inserting (5.71) into (5.72) and performing the integration over $k = \sqrt{k_x^2 + k_y^2}$ yields

$$\frac{E}{S} = \lim_{l \to 0} \frac{E(l)}{S} = -\frac{\hbar c\pi^2}{6a^3} \varsigma(-3) \tag{5.73}$$

where ς is the Riemann zeta function with $\varsigma(-3) = 1/120$. The Casimir force F per unit area acting between the plates due to the polarization of vacuum is

$$\frac{F}{S} = -\frac{d}{da}\left(\frac{E}{S}\right) = -\frac{\hbar c\pi^2}{240a^4} = -\frac{0.013}{a^4} \frac{\text{dn}}{\text{cm}^2} \tag{5.74}$$

This attractive force is very small, being equal to only 17 mV for $a = 1$ μm. Precise measurements of the force between a flat plate and a spherical lens have been performed [29]. The plate was connected to a torsion pendulum which measured the force between the two surfaces. The results of the measurements agree with the theoretical prediction to within 5%. In 2001, the Casimir force was measured for two parallel plates [30]. In spite of its small magnitude, the Casimir force has many application in nanotechnology, cosmology and microelectronics [31]. A general analysis of the Casimir effect, based upon a detailed analysis of the van der Waals force acting between two plates, has been performed by Lifshitz et al. [32]. The original idea of Casimir was to compute the van der Waals force between polarizable molecules.

5.8   Problems

Problem 5.1.

Assume that two identical particles are placed at random into one of three boxes with each box having two parts (Fig. 5.3).

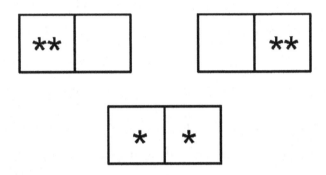

Fig. 5.3   Three boxes with each box having two parts.

Calculate the probability of finding each of distributions shown for a) classical particles, b) bosons, and c) fermions.

Solution.

a) Classical particles are distinguishable. Therefore, the lower box has two possibilities (particle 1 in the left part and particle 2 in the right part, or the opposite), and the probabilities (from left to right) are 1/4, 1/4 and 1/2.

b) Bosons are indistinguishable, and therefore, the probabilities are 1/3, 1/3 and 1/3

c) According to the Pauli principle, two fermions cannot be in the same part of the box. Therefore, the probabilities are 0, 0 and 1.

Problem 5.2.

One has to distribute three identical particles (1, 2, 3) among three states (α, β, γ). Compute the probability that all the particles will be in the same state for classical particles and for bosons [33].

Solution.

For classical particles, each of the three particles can be in each of the three states, yielding $3^3 = 27$ different possibilities. Only three of these arrangements correspond to all three particles occupying the same state (all particles in state α or in state β or in state γ). Therefore, the probability of this distribution is $3/27 = 1/9$.

For bosons, one can put all three particles in the same state (three possibilities), or two particles in the same state and the third particle in a different state $[3 \times (3 - 1)] = 6$ possibilities), or one can put each particle in a different state (one possibility). The total number of possibilities is 10. Therefore, the probability of finding all three bosons in the same state is $3/10$, which is $(3/10)/(1/9) = 2.7$ times larger than for classical particles.

Problem 5.3.

A relativistic free electron of energy E is incident on a barrier of height $V = e\phi$ located at $x = 0$. Find the reflection and transmission coefficients.

Solution.

The electron is described by the following Dirac equation,

$$(c\alpha p + \beta mc^2)\Psi = (E - V)\Psi \tag{5.75}$$

where

$$V = \left\{ \begin{array}{ll} 0, & x < 0 \\ e\phi, & x > 0 \end{array} \right\} \tag{5.76}$$

Analogous to the calculations performed in Section 5.5, the solutions of Eq. (5.75) for energy λE with $\lambda = \pm 1$, and for spin projection $\pm 1/2$ have the following form for $x < 0$

$$\Psi_{\lambda,\, p,\, 1/2} = \sqrt{\frac{mc^2 + \lambda E}{2\lambda E}} \left\{ \begin{array}{c} 1 \\ 0 \\ \frac{cp}{mc^2 + \lambda E} \\ 0 \end{array} \right\} \frac{\exp\left(ipx/\hbar\right)}{(2\pi\hbar)^{3/2}} \tag{5.77}$$

$$\Psi_{\lambda,\, p,\, -1/2} = \sqrt{\frac{mc^2 + \lambda E}{2\lambda E}} \left\{ \begin{array}{c} 1 \\ 0 \\ 0 \\ -\frac{cp}{mc^2 + \lambda E} \end{array} \right\} \frac{\exp\left(ipx/\hbar\right)}{(2\pi\hbar)^{3/2}} \tag{5.78}$$

In the region $x > 0$, one has to replace E in Eqs. (5.77)-(5.78) by $E - V$ and replace $p = (1/c)\sqrt{E^2 - m^2c^4}$ by $q = (1/c)\sqrt{(E - V)^2 - m^2c^4}$.

The spin projection is a conserved variable. Therefore, one can choose either (5.77) or (5.78). Using the former with $\lambda = 1$, the solution of the Schrödinger equation in region $x < 0$ is the sum of an incident and a reflective wave,

$$\Psi_{x<0} = A \left\{ \begin{array}{c} 1 \\ 0 \\ \frac{cp}{mc^2+\lambda E} \\ 0 \end{array} \right\} \exp\left(ipx/\hbar\right) + B \left\{ \begin{array}{c} 1 \\ 0 \\ -\frac{cp}{mc^2+\lambda E} \\ 0 \end{array} \right\} \exp\left(-ipx/\hbar\right)$$

(5.79)

and in the region $x > 0$,

$$\Psi_{x>0} = C \left\{ \begin{array}{c} 1 \\ 0 \\ \frac{cq}{mc^2+E-V} \\ 0 \end{array} \right\} \exp\left(iqx/\hbar\right) \qquad (5.80)$$

The reflection and transmission coefficients R and T are given by

$$R = \frac{|J_B|}{|J_A|}; \qquad T = \frac{|J_C|}{|J_A|} \qquad (5.81)$$

with the probability flux $J = \Psi \alpha \Psi$. Inserting this formula and (5.79)-(5.80) into (5.81), one obtains

$$R = \left| \frac{a-b}{a+b} \right|^2; \qquad T = \frac{4\,|ab|}{|a+b|}; \qquad (5.82)$$

$$a \equiv \left(\frac{E-mc^2}{E+mc^2} \right)^{1/2}; \qquad b \equiv \left(\frac{E-mc^2-V}{E+mc^2-V} \right)^{1/2}$$

Consider the following special cases [4].

a) For $V = 0$ there is no reflection.

b) For $0 < V < E - mc^2$, $R = [(a-b)/(a+b)]^2 < 1$, $T = 4ab/(a+b)^2 < 1$. In this case, the kinetic energy of the incident electron is greater than the height of the barrier, and, as in the non-relativistic case, an electron may either pass over the barrier or be reflected.

c) For $V = E - mc^2$, $R = 1$, $T = 0$. The kinetic energy of the incident electron is equal to the barrier height, and, as in the non-relativistic case, the electron is totally reflected by the barrier.

d) For $E - mc^2 < V < E + mc^2$, $R = 1$. In this case, as in the previous case, the electron is totally reflected, but, in contrast to the previous case, the electron penetrates the region $x > 0$, and, after travelling some distance in this region, returns to region $x < 0$.

e) For $E + mc^2 < V < \infty$, a new phenomenon appears. Due to the polarization of the vacuum, the surprising result $R > 1$ is obtained. This paradox (called the Klein paradox) is discussed in Section 5.7.2.

Chapter 6

Semiclassical Approximation to Quantum Mechanics

6.1 Wave equation

Just as the transition from wave optics to geometric optics occurs as the wavelength $\lambda \to 0$, the transition from quantum mechanics to classical mechanics occurs as the Planck constant $\hbar \to 0$. However, in the Schrödinger equation,

$$i\hbar \frac{\partial \Psi}{\partial t} = -\frac{\hbar^2}{2m} \nabla^2 \Psi + V\Psi \tag{6.1}$$

the small factor $\hbar^2$ multiplies the highest derivative. Therefore, it is convenient to replace the function $\Psi(\mathbf{r}, t)$ in Eq. (6.1) by a new function $W(\mathbf{r}, t)$ defined by

$$\Psi(\mathbf{r}, t) = A \exp\left(\frac{iW(\mathbf{r}, t)}{\hbar}\right) \tag{6.2}$$

Inserting (6.2) into (6.1) leads to the following equation for $W(\mathbf{r}, t)$,

$$\frac{\partial W}{\partial t} + \frac{1}{2m}(\nabla W)^2 + V - \frac{i\hbar}{2m}\nabla^2 W = 0 \tag{6.3}$$

In the limit $\hbar \to 0$, the nonlinear equation (6.3) reduces to the classical Hamilton-Jacobi equation (1.14) with $\mathbf{p} = \nabla W$. The stationary equation for the function $S(\mathbf{r})$, corresponding to energy E, can be obtained from Eq. (6.3) by the substitution $W(\mathbf{r}, t) = S(\mathbf{r}) - Et$,

$$\frac{1}{2m}(\nabla S)^2 - [E - V(\mathbf{r})] - \frac{i\hbar}{2m}\nabla^2 S = 0 \tag{6.4}$$

One can apply perturbation theory to Eq. (6.4) when the last term is small, $\hbar \nabla^2 S < (\nabla S)^2$, or equivalently,

$$p^2 > \hbar \nabla p \tag{6.5}$$

The last inequality can be rewritten by introducing the wavelength $\lambda = 2\pi\hbar/p$,

$$|\nabla\lambda| < 2\pi, \tag{6.6}$$

In other words, the wavelength has to change slowly over the characteristic length of a given system. Another way to rewrite Eq. (6.5) is by means of the classical equation of motion,

$$\frac{dp}{dx} = \frac{d}{dx}\sqrt{2m(E-V)} = -\frac{m}{p}\frac{dV}{dx} = \frac{mF}{p}, \tag{6.7}$$

which transforms Eq. (6.5) into the following form,

$$p^3 > m\hbar|F| \tag{6.8}$$

This equation shows that one cannot use perturbation theory in the vicinity of the turning point, where $V(x) = E$ and $p = 0$. For the Coulomb potential $V = -e^2/r$, one obtains the classical result $p = \sqrt{me^2/r}$, and the criterion (6.8) becomes

$$r > \frac{\hbar^2}{me^2} \equiv a_B \tag{6.9}$$

Therefore, for an atomic electron, this theory is applicable only for those orbits having a radius much larger than the Bohr radius a_B.

Let us now apply perturbation theory to the one-dimensional version of Eq. (6.4),

$$\frac{1}{2m}\left(\frac{dS}{dx}\right)^2 - (E-V) - \frac{i\hbar}{2m}\frac{d^2S}{dx^2} = 0, \tag{6.10}$$

seeking a solution of the form

$$S = S_0 + \frac{\hbar}{i}S_1 + \left(\frac{\hbar}{i}\right)^2 S_2 + \dots \tag{6.11}$$

Inserting (6.11) into (6.10) yields, in zeroth order approximation, $dS_0/dx = \pm p(x)$, implying

$$S_0(x) = \pm\int p(y)\,dy \tag{6.12}$$

The first-order perturbation theory equation is

$$\frac{d^2S_0}{dx^2} + 2\frac{dS_0}{dx}\frac{dS_1}{dx} = 0 \tag{6.13}$$

or

$$\frac{dS_1}{dx} = -\frac{d^2S_0/dx^2}{2dS_0/dx} = -\frac{1}{2p}\frac{dp}{dx} \tag{6.14}$$

yielding

$$S_1 = -\frac{1}{2} \ln p \tag{6.15}$$

Combining these two results with equation (6.2) gives the following approximate stationary solution,

$$
\Psi(x) = \frac{C_1}{\sqrt{p}} \exp\left[\frac{i}{\hbar} \int p(y)\, dy\right] \\
+ \frac{C_2}{\sqrt{p}} \exp\left[-\frac{i}{\hbar} \int p(y)\, dy\right]
\tag{6.16}
$$

where the constants of integration C_1 and C_2 are determined by the boundary conditions.

The presence of $\sqrt{p}$ in the denominator of Ψ or $1/p$ in $|\Psi|^2\, dx$ - the probability for the particle to be in the interval $(x,\ x + dx)$ - is natural for a "classical" particle. The larger the momentum, the less time the particle will spend in any given interval.

6.2 Turning points

Thus far, it was assumed that $E > V(x)$ to satisfy the classical relation $E = p^2/2m + V(x)$. However, due to the uncertainly principle, this relation is relaxed, and one has to consider $E < V(x)$ as well. Then, p becomes imaginary, $p = i\kappa$, and Eq. (6.16) takes the following form,

$$
\Psi(x) = \frac{C_3}{2\sqrt{|p|}} \exp\left[\frac{1}{\hbar} \int_a^x \kappa(y)\, dy\right] \\
+ \frac{C_4}{2\sqrt{|p|}} \exp\left[-\frac{1}{\hbar} \int_a^x \kappa(y)\, dy\right]
\tag{6.17}
$$

The typical form of the potential energy is shown in Fig. 6.1, with turning point $x = a$, for which $V(a) = E$.

The semiclassical wave function has the form (6.17) for $E < V(x)$. According to Eq. (6.8), the solutions for $E < V(x)$ and $E > V(x)$ are unreliable in the vicinity of $x = a$. To obtain the complete solution, one has to match these two equations. Starting from $x \gg a$, where one has the decreasing solution described by the second term in Eq. (6.17), one can formally avoid the point $x = a$ by performing the integration over the upper half circle of the complex x-plane, and the result of this integration has to coincide with the second term in solution (6.16) for $x < a$. Since

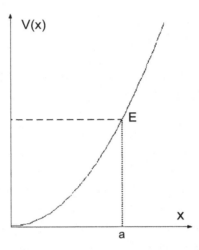

Fig. 6.1   Potential energy of the particle having energy E.

$V(x) - E \propto |x - a|$ remains unchanged along the integration path, while the phase increases by $\exp(i\pi)$ or $\sqrt{p}$ increases by $\exp(i\pi/4)$, the coefficient C_4 transforms into coefficient $C_2 = (C_4/2)\exp(-i\pi/4)$. Similarly, bypassing the point $x = a$ around the lower half circle of the complex x-plane, one gets $C_1 = (C_4/2)\exp(i\pi/4)$. Therefore, the wave function (6.17) for $x > a$ evolves into the function

$$\Psi = \frac{C_4}{\sqrt{p}} \cos\left[\frac{1}{\hbar}\int_a^x p(y)\,dy + \frac{\pi}{4}\right]$$

$$= \frac{C_4}{\sqrt{p}} \sin\left[\frac{1}{\hbar}\int_x^a p(y)\,dy + \frac{\pi}{4}\right]$$

(6.18)

for $x < a$.

6.3   Energy spectrum

The results obtained in the previous section allow us to find the energy spectrum of a particle located in different types of potentials $V(x)$. Let us extend the potential shown in Fig. 6.1 to the form depicted in Fig. 6.2, with two turning points $x = b$ and $x = a$.

The wave function for $x < a$ is given by Eq. (6.18). Analogously, one can find the wave function in the region $x > b$,

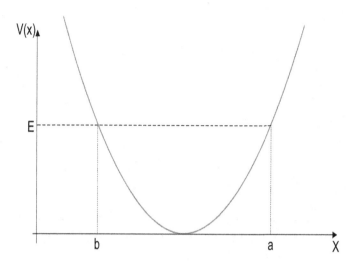

Fig. 6.2   Potential energy of a particle having energy E with two turning points.

$$\Psi = \frac{C_2'}{\sqrt{p}} \sin\left[\frac{1}{\hbar}\int_b^x p(y)\,dy + \frac{\pi}{4}\right] \tag{6.19}$$

The two functions (6.18) and (6.19) must be identical in the region of overlap, $b < x < a$. This is possible only if the sum of the arguments of the sine functions equals $\pi(n+1)$ for integer n and the amplitudes satisfy the condition $C_2' = (-1)^n C_2$,

$$\frac{1}{\hbar}\int_b^a p(y)\,dy + \frac{\pi}{2} = (n+1)\pi \tag{6.20}$$

or

$$\oint p(y)\,dy = 2\pi\hbar\left(n + \frac{1}{2}\right) \tag{6.21}$$

The last equation is the well-known Bohr-Sommerfeld quantization condition.

Using Eq. (6.21), one can find the energy spectrum for an arbitrary potential $V(x)$. As an example, let us consider a harmonic oscillator, $V(x) = kx^2/2$. The turning points are determined by the condition $p = 0$ or $x_0 = \pm\sqrt{2E/k}$. Therefore, Eq. (6.20) gives

$$\sqrt{2m}\int_{-x_0}^{x_0}\sqrt{E_n - \frac{ky^2}{2}}\,dy = \pi\hbar\left(n + \frac{1}{2}\right), \tag{6.22}$$

which, upon integration, yields the correct result,

$$E_n = \hbar\omega\left(n + \frac{1}{2}\right) \tag{6.23}$$

6.4 Tunneling through a potential barrier

The procedure of matching the wave functions at both sides of a turning point will now be used to analyze a particle passing through a potential barrier. Consider the general form of the potential barrier shown in Fig. 6.3, starting from the case when the energy of the incident particle E is less than the barrier height, $E < V_0$.

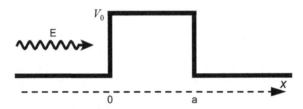

Fig. 6.3   Tunneling of a particle with energy E through a potential barrier.

The wave functions of the particle in the each of the three regions shown in Fig. 6.3 are

$$\Psi_1 = A \exp\left(ik_0 x\right) + R \exp\left(-ik_0 x\right); \qquad x < 0 \tag{6.24}$$

$$\Psi_2 = \frac{C_1}{\sqrt{\kappa}} \exp\left[\frac{1}{\hbar}\int \kappa\left(y\right) dy\right] + \frac{C_2}{\sqrt{\kappa}} \exp\left[-\frac{1}{\hbar}\int \kappa\left(y\right) dy\right]; \; 0 < x < a$$

$$\Psi_3 = T \exp\left(ikx\right); \qquad x > a$$

Region 2 is the quasi-classical region for which the wave function was calculated in the previous section. Regions 1 and 3 contain the reflected and transmitted waves. The matching of the wave functions and their derivatives at the boundary points, $x = 0$ and $x = a$, and the normalization condition gives five conditions for the five amplitudes in (6.24). In calculating the derivatives, we take into account only the larger term, arising from the differentiation of the exponents. The solutions of these equations for the transmission coefficient gives

$$\left|\frac{T}{A}\right|^2 \propto \exp\left(-2qa\right) \tag{6.25}$$

where

$$q = \sqrt{2m\left(V_0 - E\right)/\hbar^2} \tag{6.26}$$

One sees from Eq. (6.25) that the transmission coefficient, $|T/A|^2$ strongly depends on the mass of the particle. For a proton with mass 1840 times the electron mass, the the transmission coefficient between the two particles differs by the factor $\exp\left(\sqrt{1840}\right) \approx 10^{18}$!

If the energy of the incident particle E is larger than the barrier height, $E > V_0$, a similar calculation for the reflecting coefficient leads to the following result,

$$\left|\frac{R}{A}\right|^2 = \frac{V_0^2 \sin^2(qa/\hbar)}{V_0^2 \sin^2(qa/\hbar) + 4E(E - V_0)} \qquad (6.27)$$

For $qa/\hbar = \pi n$ with integer n, the reflection coefficient $|R/A|^2$ vanishes, implying that the particle does not feel the presence of the barrier. For the barrier shown in Fig. 6.3, this effect occurs when the energy of the incident particle E is larger than the barrier height. This interesting phenomenon of resonance tunneling also occurs for a particle moving through the two equal rectangular potential barriers shown in Fig. 6.4

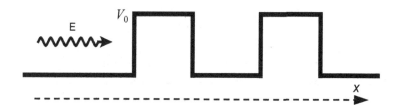

Fig. 6.4   Tunneling through a double potential barrier.

For a special energy of the incident particle, the transmission coefficient becomes nearly unity even though each of the barriers by itself has a low transparency. This phenomenon of resonance takes place when the energy E is equal to the bound state energy corresponding to this potential. The same phenomenon occurs for a particle tunneling through many symmetrically located potential barriers [34]. This phenomenon bears a close resemblance to the Bloch theorem in solids, according to which an electron moves without resistance in a periodic field of ions forming a crystalline lattice.

6.5 Problems

Problem 6.1.
   Using the semiclassical approximation, consider the energy levels of a
particle in the symmetric double-well potential $V(x)$ shown in Fig. 6.5.

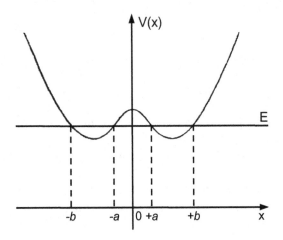

Fig. 6.5   Symmetric double-well potential.

   a) Evaluate the energy levels in each potential well.
   b) Find the energy levels in the double-well potential $V(x)$.
Solution.
   a) The energy levels E_n^0, which correspond to the wave functions $\psi_{n,0}$,
are determined by the Bohr-Sommerfeld quantization conditions,

$$\int_a^b \sqrt{2m\left(E_n^0 - V\right)}dx = \pi\hbar\left(n + 1/2\right) \tag{6.28}$$

   b) As was done for (6.28), the energy levels can be obtained by solving
the Schrödinger equation in each of the five regions of $V(x)$, requiring
the continuity of the wave function and its derivative at each boundary
[8]. However, assuming that the transition probability through the barrier
is small, one can find the separation of the energy levels E_n into $E_{n,1}^0$
and $E_{n,2}^0$ induced by these transitions and described by the symmetric and

antisymmetric combinations of the functions $\psi_{n,0}(x)$ and $\psi_{n,0}(-x)$,

$$\psi_{n,1}(x) = \frac{1}{\sqrt{2}}\left[\psi_{n,0}(x) + \psi_{n,0}(-x)\right], \tag{6.29}$$

$$\psi_{n,2}(x) = \frac{1}{\sqrt{2}}\left[\psi_{n,0}(x) - \psi_{n,0}(-x)\right]$$

Because of the exponential decay of the functions $\psi_{n,0}(\pm x)$ beyond the wells, one can assume that each of these two functions is located within its own well. Multiplying the Schrödinger equations

$$\frac{d^2\psi_{n,0}}{dx^2} + \frac{2m}{\hbar^2}(E_{n,0} - V)\psi_{n,0} = 0, \tag{6.30}$$

$$\frac{d^2\psi_{n,1}}{dx^2} + \frac{2m}{\hbar^2}(E_{n,1} - V)\psi_{n,1} = 0$$

by $\psi_{n,1}$ and $\psi_{n,0}$, respectively, subtracting term by term, and integrating from zero to infinity, one obtains,

$$E_{n,1} - E_{n.0} = -\frac{\hbar^2}{2m}\psi_{n,0}(0)\frac{d\psi_{n,0}}{dx}(0) \tag{6.31}$$

Equation (6.31) was obtained assuming that at $x = 0$, $\psi_{n,1} = \sqrt{2}\psi_{n,0}$, $d\psi_{n,1}/dx = 0$, and

$$\int_0^\infty \psi_{n,0}\,\psi_{n,1}dx \approx \frac{1}{\sqrt{2}} \tag{6.32}$$

Analogously, one can find $E_{n,2}-E_{n.0}$, which, together with (6.31), yields

$$E_{n,2} - E_{n,1} = \frac{2\hbar^2}{m}\psi_{n,0}(0)\frac{d\psi_{n,0}}{dx}(0) \tag{6.33}$$

$$= \frac{\omega\hbar}{\pi}\exp\left(-\frac{1}{\hbar}\int_{-a}^{a}|p|\,dx\right)$$

where ω is the classical frequency in a single well, $\omega^{-1} = (m/\pi)\int_a^b dx/p$. The explicit form of the semiclassical wave function [35] has been used in the last equality in (6.33).

Problem 6.2. The extension of the previous problem to the case of N equal potential wells shown in Fig. 6.6 requires quite cumbersome calculations [8], and we therefore restrict ourselves to the final results. The Bohr-Sommerfeld quantization conditions are

$$\int_{a_1}^{b_1} pdx = \pi\hbar\left(n + \frac{1}{2}\right) \tag{6.34}$$

$$+\hbar\cos\left(\frac{\pi s}{N+1}\right)\exp\left(-\frac{1}{\hbar}\int_{b_1}^{a_2}|p|\,dx\right)$$

where $n = 0, 1, 2, ...$ and $s = 1, 2, 3, ...$The latter equation is similar to Eq. (6.28) of the previous problem, which gives the energy levels in a single well. Also analogous to the previous problem, the energy level of a single well is split into N sublevels,

$$\Delta E_n = \frac{\hbar \omega}{\pi} \cos \left(\frac{\pi s}{N+1} \right) \exp \left(-\frac{1}{\hbar} \int_{b_1}^{a_2} |p| \, dx \right) \qquad (6.35)$$

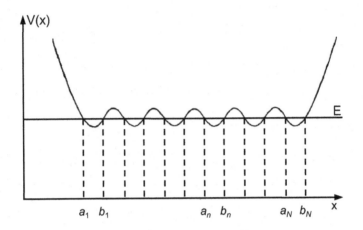

Fig. 6.6   Potential $V(x)$ consisting of many potential wells.

Problem 6.3. In the semiclassical approximation, find the form of a monotonically increasing symmetric potential $V(x) = V(-x)$ that yields a discrete energy spectrum $E = E(n)$. As an example, consider the harmonic oscillator for which $V = m\omega^2 x^2/2$.

Solution.

The Bohr-Sommerfeld quantization condition for a symmetric potential

$$\pi \hbar \left(n + \frac{1}{2} \right) = 2 \int_0^a \sqrt{2m \left[E - V(x) \right]} \, dx \qquad (6.36)$$

may be viewed as an integral equation for the function $V(x)$. Differentiating Eq. (6.36) with respect to E and considering V (with $V(a) = E$) as the independent variable, one obtains

$$\frac{\pi \hbar}{\sqrt{2m}} \frac{dn}{dE} = \int_0^E \frac{dx}{dV} \frac{dV}{\sqrt{E-V}} \qquad (6.37)$$

Multiplying by $(\beta - E)^{-1/2}$ and integrating over E from zero to β yields

$$\frac{\pi\hbar}{\sqrt{2m}} \int_0^\beta \frac{dn}{dE} \frac{dE}{\sqrt{(\beta - E)}} = \int_0^\beta dE \int_0^E \frac{dx}{dV} \frac{dV}{\sqrt{(\beta - E)(E - V)}} =$$

$$\int_0^\beta \frac{dx}{dV} dV \int_0^\beta \frac{dE}{\sqrt{(\beta - E)(E - V)}} = \pi x(\beta) \qquad (6.38)$$

Solving for $x(\beta)$ and setting $\beta = V$ yields

$$x(V) = \frac{\hbar}{\sqrt{2m}} \int_0^V \frac{dE}{(dE/dn)\sqrt{V - E}} \qquad (6.39)$$

The derivative dE/dn can be found from the function $E(n)$. Inverting from $x(V)$ to $V(x)$ then solves the problem.

For the harmonic oscillator, $E(n) = \hbar\omega(n + 1/2)$ so that $dE/dn = \hbar\omega$ and Eq. (6.39) gives $x = \sqrt{2V/m\omega^2}$, which yields the correct answer $V = m\omega^2 x^2/2$.

Chapter 7

Scattering

Scattering is one of the most effective experimental techniques for studying Nature. One of the most dramatic events in experimental physics was the Rutherford experiment of scattering alpha particles by atoms of gold. In 1911 he found that the most of the particles passed freely through the atom, and only a small percentage was scattered, which proved that the atom is "empty" since the radius of the nucleus (10^{-12} cm) is 10,000 times smaller than the radius of the atom (10^{-8} cm). It was the entrance into the new quantum world because the radius of an atom is a million times smaller than the thickness of a human hair.

After describing the phenomenological characteristics of scattering in Section 7.1, we shall consider the two main theoretical descriptions of scattering, the Born approximation of perturbation theory and the method of partial waves. The former is based on the properties of the scatterer and describes the scattering of high-energy particles, whereas the latter is the basis for understanding the properties of incident low-energy particles.

7.1 Phenomenological description of scattering

The two asymptotic forms of the quantum mechanical description of the scattering process are the incident and outgoing wave functions. The incident plane wave has the form $\exp(ikz)$, where z is chosen in the direction of the incident particle. According the Huygens principle in optics, all points of a wave front may be regarded as new sources of waves. Therefore, one can describe the outgoing wave at a large distance from the scatterer as a spherical wave, $f(\theta, \phi) \exp(ikr)/r$. The scattering is usually symmetric with respect to the z-axis, and, therefore, independent of the angle ϕ. Combining these two results, one can write the approximate form of the wave function of a scattered particle as

$$\Psi = \exp\left(ik_{in}z\right) + f\left(\theta\right)\frac{\exp\left(ik_{out}r\right)}{r} \tag{7.1}$$

where $f\left(\theta\right)$ is called the scattering amplitude.

The incident particle flux density J_{in} and the outgoing (scattered) radial flux density J_{out} can be found from Eq. (5.1)

$$J_{in} = \frac{\hbar k_{in}}{m}; \qquad J_{out} = \left|f\left(\theta\right)\right|^2\frac{\hbar k_{out}}{mr^2} \tag{7.2}$$

Let us define the scattering cross-section σ. In classical mechanics, the cross-section of the scattering by a spherical object of radius r is equal to πr^2. The question is whether this result is retained in quantum mechanics. The differential cross-section $d\sigma\left(\theta\right)$ is defined as the number of particles $dN\left(\theta,\phi\right)$ scattered into solid angle $r^2 d\Omega$, $dN\left(\theta,\phi\right) = J_{out}r^2 d\Omega$, relative to the current J_{in} of the incident particles,

$$\frac{d\sigma\left(\theta\right)}{d\Omega} = \frac{J_{out}r^2}{J_{in}} = \left|f\left(\theta\right)\right|^2 \tag{7.3}$$

In the last equation, we assume $k_{out} = k_{in}$, which restricts our consideration to elastic scattering. The total scattering cross-section is given by

$$\sigma = \int \frac{d\sigma}{d\Omega}d\Omega = 2\pi \int \left|f\left(\theta\right)\right|^2 \sin\theta d\theta \tag{7.4}$$

Hence, the quantum mechanical scattering is determined by the scattering amplitude $f\left(\theta\right)$.

7.2 Born approximation

In section 4.2, we showed that the solution of the Schrödinger equation of a particle moving in the spherically symmetric potential $V\left(r\right)$

$$\left[-\frac{\hbar^2}{2m}\nabla^2 + V\left(r\right)\right]\Psi = E\Psi \tag{7.5}$$

can be written in terms of the Green function, defined in (4.12). In our case, the Green function is $-\exp\left(ikr\right)/4\pi r$, and the differential equation (7.5) reduces to an integral equation of the form

$$\Psi\left(\mathbf{r}\right) = \Psi_0\left(\mathbf{r}\right) - \frac{m}{2\pi\hbar^2}\int \frac{\exp\left[ik\left|\mathbf{r}-\mathbf{r}_1\right|\right]}{\left|\mathbf{r}-\mathbf{r}_1\right|}V\left(r_1\right)\Psi\left(\mathbf{r}_1\right)d\mathbf{r}_1 \tag{7.6}$$

where $\Psi_0\left(\mathbf{r}\right) = \exp\left(i\mathbf{kr}\right)$ is the incident wave.

At this stage, we assume that the potential $V(r)$ is localized in space. Therefore, in the asymptotic region $r >> r_1$, one can neglect r_1 in the denominator of Eq. (7.6) and expand the $|\mathbf{r} - \mathbf{r}_1|$ in the exponent in a series,

$$|\mathbf{r} - \mathbf{r}_1| = \sqrt{(\mathbf{r} - \mathbf{r}_1)^2} \approx \sqrt{r^2 - 2\mathbf{r}\mathbf{r}_1} \approx r - \frac{\mathbf{r}}{r}\mathbf{r}_1 = r - \hat{\mathbf{r}}\mathbf{r}_1 \qquad (7.7)$$

where $\hat{\mathbf{r}}$ is a unit vector in the $\mathbf{r}$-direction. Therefore, the asymptotic form of Eq. (7.6) is the following,

$$\Psi(\mathbf{r}) = \Psi_0(\mathbf{r}) - \frac{\exp(ikr)}{r}\frac{m}{2\pi\hbar^2}\int \exp(-i\mathbf{k}_1\mathbf{r}_1)V(r_1)\Psi(\mathbf{r}_1)\,d\mathbf{r}_1 \qquad (7.8)$$

where $\mathbf{k}_1 = k\hat{\mathbf{r}}$ is the wave vector of the outgoing wave.

The integral equation (7.8) can be solved by perturbation theory. In first-order perturbation theory, called the Born approximation, one replaces the function $\Psi(\mathbf{r}_1)$ by the function Ψ_0 from the zero-order approximation,

$$\Psi(\mathbf{r}) = \Psi_0(\mathbf{r}) - \frac{\exp(ikr)}{r}\frac{m}{2\pi\hbar^2}\int \exp[-i(\mathbf{k}_1 - \mathbf{k})\mathbf{r}_1]V(r_1)\,d\mathbf{r}_1 \qquad (7.9)$$

Comparing (7.9) and (7.1) yields the scattering amplitude,

$$f(\theta) = -\frac{m}{2\pi\hbar^2}\int \exp(-i\mathbf{q}\mathbf{r}_1)V(r_1)\,d\mathbf{r}_1 \qquad (7.10)$$

where $\mathbf{q} = \mathbf{k}_1 - \mathbf{k}$ is the change of the wave vector of a particle due to elastic scattering $(k_1 = k)$,

$$q^2 = (\mathbf{k}_1 - \mathbf{k})^2 = 2k^2 - 2k^2\cos\theta = 4k^2\sin^2\frac{\theta}{2} \qquad (7.11)$$

According to Eq. (7.10), the particle is scattered by the Fourier component of the scattered potential.

The criterion for the applicability of the Born approximation is the smallness of the function obtained in the first approximation compared with the function in the zero approximation. Consider the Born approximation for a spherically-symmetric potential $V(r)$ of magnitude V_0 confined in a region $r < a$. For a low energy incident particle, $ka << 1$, one can replace the exponent in the integral of Eq. (7.10) by unity. Evaluating of our estimation at $r = 0$ yields for the application of the Born approximation,

$$\frac{\hbar^2}{2ma^2} > V_0 \qquad (7.12)$$

In other words, the kinetic energy $\hbar^2/2ma^2$ of the incident particle, confined to a volume equal to the range of the potential, must be larger than the potential energy of the particle. Since $\hbar^2/2ma^2$ is the minimum depth of a potential well that contains a bound state, the physical meaning of criterion (7.12) is that the interaction energy has to be smaller than the potential depth that produces a bound state. The Born approximation works even better at higher energies, where the oscillatory parts of the exponential substantially decrease the value of the integral in Eq. (7.6).

7.3   Scattering by different potentials

Consider the scattering by a periodic lattice with period $\mathbf{R}$, $V(\mathbf{r}) = V(\mathbf{r} + \mathbf{R})$. Upon a change of variable from $\mathbf{r}$ to $\mathbf{r} + \mathbf{R}$, the scattering amplitude (7.10) becomes

$$f(\theta) = -\frac{m}{2\pi\hbar^2} \int \exp\left[-i\mathbf{q}\left(\mathbf{r}_1 + \mathbf{R}\right)\right] V(\mathbf{r}_1)\, d\mathbf{r}_1 \qquad (7.13)$$

and, by substructing (7.13) from (7.10),

$$\frac{m}{2\pi\hbar^2} \int \exp\left(-i\mathbf{q}\mathbf{r}_1\right)\left[1 - \exp\left(-i\mathbf{q}\mathbf{R}\right)\right] V(\mathbf{r}_1)\, d\mathbf{r}_1 = 0, \qquad (7.14)$$

which shows that $\mathbf{q}\mathbf{R} = 2\pi n$ with integer n. This condition means that the scattering on a lattice occurs when the scattering vector $\mathbf{q} = \mathbf{k}_1 - \mathbf{k}$ coincides with a vector of the reciprocal lattice (Bragg condition).

Let us now compare the results of the Born approximation for three different potentials; the Yukawa potential, $V_{yu} = V_0 \exp\left(-\mu r\right)/r$, the Gaussian potential, $V_G = V_0 \exp\left(-r^2/4a^2\right)$, and the square-well potential $V_{sq} = V_0$ at $r < a$ and $V_{sq} = 0$ at $r > a$.

The Yukawa potential is just the Coulomb potential screened at length μ^{-1}. Inserting into Eq. (7.10) and integration over θ and r_1 gives

$$f(\theta) = -\frac{2mV_0}{\hbar^2\left(q^2 + \mu^2\right)} \qquad (7.15)$$

and

$$\frac{d\sigma}{d\Omega} = |f(\theta)|^2 = \frac{4m^2 V_0^2}{\hbar^4\left(q^2 + \mu^2\right)^2} \qquad (7.16)$$

For the Coulomb potential, $\mu = 0$, and Eq. (7.16) reduces to the Rutherford formula, since $q = 2k\sin\left(\theta/2\right)$,

$$\frac{d\sigma}{d\Omega} = \frac{m^2 V_0^2}{4\hbar^4 k^4 \sin^4\left(\theta/2\right)} \qquad (7.17)$$

For the Gaussian potential,

$$f\left(\theta\right) = -\frac{2mV_0}{\hbar^2 q} \int_0^\infty \sin\left(qr_1\right) \exp\left(-r_1^2/4a^2\right) r_1 dr_1$$

$$= \frac{\sqrt{2\pi} mV_0 a^3}{\hbar^2} \exp\left(-q^2 a^2\right) \qquad (7.18)$$

Finally, for the square-well potential,

$$f\left(\theta\right) = -\frac{mV_0}{2\pi\hbar^2} \int_0^a r_1 \sin\left(qr_1\right) dr_1 \qquad (7.19)$$

$$= \frac{2mV_0}{q^3\hbar^2} \left[qa\cos\left(qa\right) - \sin\left(qa\right)\right]$$

Comparing (7.17), (7.18) and (7.19) shows that in each case, the differential cross-section for small aq is proportional to $(aq)^2$, although for large aq, the results are quite different for each case (Fig. 7.1). Therefore, experiments at small aq do not enable one to determine the form of the scattering potential.

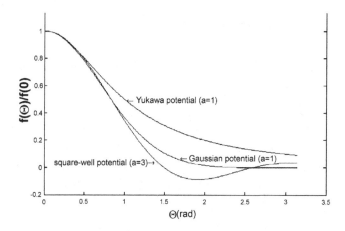

Fig. 7.1 Scattering amplitude as a function of aq for the screened Coulomb, Gaussian and square-well potentials.

7.4 Partial waves

In analyzing the Klein paradox in Section 5.7.2, we saw that under special conditions, the particle undergoes total reflection from the potential barrier, i.e. the amplitudes of the incident and reflected waves are equal. However, the two waves may have different phases. Such a phase difference in the scattering process provides the basis for the method of partial waves. The solution of the Schrödinger equation in a centrally symmetrical potential $V(r)$ can be written as the sum of the products of the radial function $R_l(r) = \chi_l(r)/r$ and the Legendre polynomials $P_l(\cos\theta)$. The differential equation for the function $\chi_l(r)$ has the following form [35],

$$\frac{d^2\chi_l}{dr^2} + \left[k^2 - \frac{l(l+1)}{r^2} - \frac{2m}{\hbar^2}V(r)\right]\chi_l = 0; \quad k^2 = \frac{2mE}{\hbar^2} \tag{7.20}$$

For a restricted potential $V(r)$, the asymptotic $(r \to \infty)$ solution of the second-order differential equation (7.20) can be written as

$$\chi_l(r) = A\sin(kr + \delta_l), \tag{7.21}$$

where the phase shift δ_l is determined from the boundary conditions. One can express the total cross-section σ for the scattering [35] in terms of the phase shifts δ_l,

$$\sigma = \frac{4\pi}{k^2}\sum_l (2l+1)\sin^2\delta_l, \tag{7.22}$$

For s-wave scattering $(l = 0)$, Eq. (7.22) reduces to

$$\sigma = \frac{4\pi}{k^2}\sin^2\delta_0 \tag{7.23}$$

As an example, consider $s-$wave scattering by a hard sphere,

$$V(r) = 0, \; r > a; \qquad V(r) = \infty, \; r \le a \tag{7.24}$$

In the region $r \le a$, $\chi(r) = 0$. Therefore, the boundary condition at $r = a$ for the function (7.21) leads to $\delta_0 = -ka$. According to Eq. (7.23), for $ka \ll 1$, one obtains

$$\sigma = \frac{4\pi}{k^2}\sin^2\delta_0 \approx \frac{4\pi}{k^2}\delta_0^2 = 4\pi a^2 \tag{7.25}$$

This quantum mechanical result is four times larger than the classical result.

We now consider the following potential,

$$V(r) = 0, \; r > a; \qquad V(r) = -V_0, \; r \le a \tag{7.26}$$

Unlike the previous problem, one has to solve the Schrödinger equation in both regions $r < a$ and $r > a$, and then to equate the functions and their derivatives at the boundary $r = a$. The solution for $r > a$ is given by equation (7.21) whereas the solution for $r < a$ is $\chi = B \sin(k_0 r)$, where $k_0 = \left[2m(E + V_0)/\hbar^2\right]^{1/2}$.

Since we are interested in the phase shift δ_0, it is convenient to require continuity of the ratio $\chi/(d\chi/dr)$ at the boundary $r = a$,

$$k_0 \tan(ka + \delta_0) = k \tan(k_0 a) \tag{7.27}$$

or

$$\tan \delta_0 = \frac{(k/k_0) \tan(k_0 a) - \tan(ka)}{1 + (k/k_0) \tan(k_0 a) \tan(ka)} \tag{7.28}$$

For low energies, $ka << 1$,

$$\tan \delta_0 = ka \left(\frac{\tan(k_0 a)}{ak_0} - 1\right) \tag{7.29}$$

Inserting (7.29) into (7.23) yields for $ka << 1$,

$$\sigma = 4\pi a^2 \left(\frac{\tan(k_0 a)}{ak_0} - 1\right)^2 \tag{7.30}$$

In the case of a potential barrier $V = +V_0$, instead of the potential well $V = -V_0$, one has to replace $\tan(k_0 a)$ in Eq. (7.30) by $\tanh(k_0 a)$.

Interesting conclusions can be drawn from Eq. (7.28). For a special relation between the energy E of the particle and the characteristics V_0 and a of the scatterer, so that $\tan(k_0 a)/k_0 \approx \tan(ka)/k$ and $\tan \delta_0 \approx 0$, the particle is not scattered. This is called the Ramsauer effect, which was discovered by Ramsauer in experiments of electron scattering by noble gas atoms. On the other hand, for $k_0 a$ or ka close to $\pi/2$, $\tan \delta_0$ and σ rise steeply. This condition for strong scattering coincides with the appearance of an energy level in the potential well, which "attracts" an incident electron to stay in this well. In such a manner, by a continuous change of the energy of the incoming electron, one moves back and forth from weak to strong scattering.

7.5 Optical theorem

The optical theorem connects the total cross-section of the scattering with the intensity of the forward scattering amplitude. The optical theorem has a fascinating history, being rediscovered many times [36]. Starting with the wave function Ψ for elastic scattering with $k_{in} = k_{out} = k$, one can find the intensity of the wave, which is proportional to $|\Psi|^2$,

$$|\Psi|^2 \approx 1 + f(\theta) \frac{\exp(ikr)}{r} \exp(-ikz) \tag{7.31}$$
$$+ f^*(\theta) \frac{\exp(-ikr)}{r} \exp(ikz)$$

where we neglect terms proportional to $1/r^2$ in this asymptotic expansion. Moreover, restricting consideration to very small scattering angles, $\theta \ll 1$, one may replace $f(\theta)$ and $f^*(\theta)$ by $f(0)$ and $f^*(0)$, r by $r = z/\cos\theta \approx z$, and

$$r = \sqrt{x^2 + y^2 + z^2} \approx z + \frac{x^2 + y^2}{2z} \tag{7.32}$$

Equation (7.31) can be rewritten as

$$|\Psi|^2 = 1 + \frac{f(0)}{z} \exp\left[ik(x^2 + y^2)/2z)\right]$$
$$+ \frac{f^*(0)}{z} \exp\left[-ik(x^2 + y^2)/2z\right] \tag{7.33}$$
$$= 1 + 2\operatorname{Re}\frac{f(0)}{z} \exp\left[ik(x^2 + y^2)/2z\right]$$

The total energy collected on a screen of radius R located in the $x - y$ plane is obtained by integrating Eq. (7.33) over the circle $x^2 + y^2 = R^2$. We assume that R is large enough that $kR \gg 2\pi$, but still sufficiently small that $R \ll z$, which allows one to assume small angles $\theta \ll 1$. Then,

$$\int |\Psi|^2 \, dS = \pi R^2 + \tag{7.34}$$
$$2\operatorname{Re}\frac{f(0)}{z} \int \exp\left(ikx^2/2z\right) dx \int \exp\left(iky^2/2z\right) dy =$$
$$= \pi R^2 + \frac{4\pi}{k} \operatorname{Im} f(0)$$

where the limits of integration were extended to infinity. The optical theorem (7.34) relates the imaginary part of the forward scattering amplitude to the total cross-section. It tells us that $f(\theta)$ cannot be real in all directions. In particular, f has a positive imaginary part in the forward direction.

7.6   Problems

Problem 7.1.

a) Find the conditions for the validity of the Born approximation in scattering theory.

b) In the Born approximation, find the scattering amplitude and the total cross-section for the exponential potential $V(r) = V_0 \exp(-r/R_0)$ in the limit of small K ($KR_0 \ll 1$) and large K ($KR_0 \gg 1$), where $\mathbf{K} = \mathbf{k}_0 - \mathbf{k}_1$ is the scattering vector.

c) Find the phase shift in the Born approximation.

d) Use the answer to the previous question to find the phase shift and the total cross-section for s-wave scattering ($l = 0$) for the exponential potential and compare the results obtained with those obtained in part (b).

e) Experiments of the scattering of electrons by protons in atomic hydrogen show that the proton has an "exponential" charge density $\rho(r) = \rho_0 \exp(-r/R_0)$. Find the constants ρ_0 and R_0 such that the proton charge is equals e, the charge of the electron, and show that the proton mean square radius $\langle r^2 \rangle$ is equal to $12R_0^2$ [33].

Solution.

a) Solving the Schrödinger equation for a particle moving in a potential $V(r)$ yields

$$\Psi(\mathbf{r}) = \exp(ikz) - \frac{m}{2\pi\hbar^2} \int \frac{\exp(ik|\mathbf{r} - \mathbf{r}_1|)}{|\mathbf{r} - \mathbf{r}_1|} \exp(ikz_1) V(r_1) d\mathbf{r}_1 \quad (7.35)$$

For the rapidly decreasing function $V(r_1)$, which is non-zero only in a region $R_0 \ll r_1$, one can transform Eq. (7.35) in the asymptotic regime $r > r_1$ into the following form,

$$\Psi(r)_{r\to\infty} \to \exp(ikz) \quad (7.36)$$

$$-\frac{m\exp(ikr)}{2\pi\hbar^2 r} \int V(r_1) \exp\{ik[z_1 - \cos(\mathbf{r}\mathbf{r}_1)r_1]\} d\mathbf{r}_1$$

which can be rewritten as

$$\Psi(\mathbf{r}) = \Psi_0(\mathbf{r}) - \frac{\exp(ikr)m}{2\pi r\hbar^2} \int \Psi_0(r_1) V(r_1)\Psi(r) d\mathbf{r}_1 \quad (7.37)$$

which is an integral equation for the function $\Psi(\mathbf{r})$, and hence equivalent to the original Schrödinger equation.

Comparing Eq. (7.36) with the asymptotic solution of the Schrödinger equation,

$$u(r) = \exp(ikz) + f(\theta, \phi) \frac{\exp(ikr)}{r} \quad (7.38)$$

yields the scattering amplitude

$$f(\theta, \phi) = -\frac{m}{2\pi\hbar^2} \int V(r_1) \exp(i\mathbf{Kr}_1) d\mathbf{r}_1 \qquad (7.39)$$

For elastic scattering $|\mathbf{k}_0| = |\mathbf{k}_1|$, the vectors $\mathbf{k}_0$ and $\mathbf{k}_1$ are oriented along the incident and scattered beam, respectively. According to Eq. (7.39), the scattering amplitude is proportional to the Fourier transform of the potential. Since the cross-section for scattering $d\sigma/d\Omega$ is defined through the scattering amplitude, $d\sigma/d\Omega = |f(\theta, \phi)|^2$, the total cross-section σ is given by

$$\sigma = \int |f(\theta, \phi)|^2 \, d\Omega \qquad (7.40)$$

Equation (7.37) can be solved by perturbation theory (first Born approximation) by replacing $\Psi(\mathbf{r})$ in the integrand of (7.37) by $\Psi_0(\mathbf{r})$. This is justified if the first correction of perturbation theory is small,

$$\langle V \rangle << \frac{\hbar^2}{2mR_0^2}, \qquad (7.41)$$

where

$$\langle V \rangle = \frac{1}{4\pi R_0^2} \left| \int \frac{V(r)}{r} d\mathbf{r} \right| \qquad (7.42)$$

is the average potential energy. According to (7.41), $\langle V \rangle$ has to be smaller than the uncertainty in the kinetic energy of a particle in the region R_0 of the scattered potential. Therefore, Eq. (7.41) is the condition for the validity of the Born approximation.

b) Evaluating the integrals (7.39) and (7.40), one obtains

$$f = -\frac{4mV_0R_0^3}{\hbar^2\left(1 + K^2R_0^2\right)^2}; \qquad \sigma \approx f^2 = \frac{16m^2V_0^2R_0^6}{\hbar^4\left(1 + K^2R_0^2\right)^4} \qquad (7.43)$$

which takes the following form in the limiting cases of small and large energies,

$$\sigma_{E\to 0} \approx V_0^2 R_0^6 \qquad \sigma_{E\to\infty} \approx \frac{V_0^2}{R_0^2 K^8} \qquad (7.44)$$

c) A plane wave can be written as a sum of spherical waves

$$\exp(ikz) = \exp(ikr\cos\theta) = \sum_l i^l(2l+1)j_l(kr)P_l(\cos\theta) \qquad (7.45)$$

where $j_l(kr)$ is the spherical Bessel function.

From this expansion, one obtains the phase shift for s-wave scattering,

$$\delta_0 (k) = -\frac{\pi m}{\hbar^2} \int_0^\infty rV(r) j_{1/2}^2 (kr) dr \qquad (7.46)$$

$$= -\frac{2m}{k\hbar^2} \int_0^\infty V(r) \sin^2 (kr) dr$$

d) Inserting $V(r)$ into Eq. (7.46) and evaluating the integral yields

$$\delta_0 (k) = -\frac{4mR_0^3 V_0 k}{\hbar^2 (1 + 4k^2 R_0^2)} \qquad (7.47)$$

For $kR_0 = 1$, $\delta_0 (k) << 1$, because, according to (7.41), $|V_0| << \hbar^2/mR_0^2$. The total cross-section σ of slow particles $(kR_0 << 1)$ is then $\sigma = 4\pi \left[\delta_0 (k) / k \right]^2$.

e) The proton charge e is given by

$$e = 4\pi \int_0^\infty \rho(r) r^2 dr = 8\pi \rho_0 R_0^3 \qquad (7.48)$$

The probability of having charge density $e\rho(r)$ is determined by the square of the normalized wave function $|\psi(r)|^2$,

$$4\pi \int_0^\infty |\psi(r)|^2 r^2 dr = 1 \qquad (7.49)$$

Comparing (7.48) and (7.49), one obtains

$$|\psi(r)|^2 = \frac{\exp(-r/R_0)}{8\pi R_0^3}, \qquad (7.50)$$

which yields

$$\langle r^2 \rangle = 4\pi \int_0^\infty |\psi(r)|^2 r^4 dr = 12R_0^2 \qquad (7.51)$$

Problem 7.2.

In order to escape from a metal, an electron must overcome the potential barrier which keeps it inside the metal. The simplest model for such a barrier is an one-dimensional step potential $V(x) = -V_0$ inside a metal and $V(x) = 0$ outside. Electrons with energy $E > V_0$ are able to escape from the metal (Fig. 7.2).

a) Find the reflection coefficient for an electron at the metal-air boundary, and compare with the classical result.

b) In the simple model considered above, the potential has a jump at the metal-air boundary. However, in a real metal, this change occurs continuously over a distance of order a lattice constant. Therefore, one can improve the model by approximating the potential in the immediate

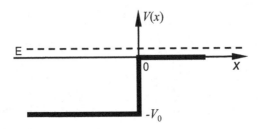

Fig. 7.2   Potential $V(r) = -V_0$ which prevents an electron from escaping from the metal.

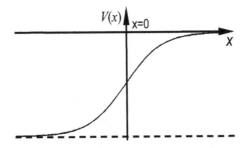

Fig. 7.3   More realistic potential $V(r) = V_0/[1 + \exp(x/a)]$ which prevents an electron from escaping from the metal.

vicinity of the surface by $V = V_0/[1 + \exp(x/a)]$ (Fig. 7.3). Find the reflection coefficient for electrons with $E > 0$ and compare with the result obtained earlier.

   Solution.

   a) $\Psi_{x<0} = C_1 \exp(ikx) + C_2 \exp(-ikx)$ with $k = \sqrt{2m(E + V_0)}/\hbar$ inside the metal and $\Psi_{x>0} = C_3 \exp\left(ix\sqrt{2mE}/\hbar\right)$ outside the metal. The continuity of the wave function and its derivatives at $x = 0$ yields the reflection coefficient R

$$R = \frac{V_0^2}{\left[\sqrt{E + V_0} + \sqrt{E}\right]^4} \qquad (7.52)$$

For $E = 0$, $R = 1$. For $E \gg V_0$ and $E \ll V_0$, one obtains, respectively,

$$R_{E \gg V_0} \approx \frac{V_0^2}{16E^2}; \qquad R_{E \ll V_0} \approx 1 - 4\sqrt{\frac{E}{V_0}} \qquad (7.53)$$

For metals, $V_0 \approx 10$ eV. Therefore, for electrons with energy $E = 0.1$ eV, $R \approx 0.67$. According to classical mechanics, an electron with $E \geq 0$ will always escape from the metal.

b) In this case, the Schrödinger equation has the following form

$$\frac{\hbar^2}{2m}\frac{d^2\Psi}{dx^2} + \left(E + \frac{V_0}{1 + \exp(x/a)}\right)\Psi = 0 \qquad (7.54)$$

Changing variables,

$$\xi = -\exp(-x/a); \qquad \Psi = \xi^{ika}\Phi(\xi); \qquad k = \frac{\sqrt{2mE}}{\hbar} \qquad (7.55)$$

yields the hypergeometric equation for the function $\Phi(\xi)$,

$$\xi(1-\xi)\frac{d^2\Phi}{d\xi^2} + (1-\xi)(1-2ika)\frac{d\Phi}{d\xi} - \frac{2ma^2V_0}{\hbar^2}\Phi = 0 \qquad (7.56)$$

Using the properties of the hypergeometric functions [10], one obtains [8],

$$R = \frac{\sinh^2\left[\pi a\left(\kappa - k\right)\right]}{\sinh^2\left[\pi a\left(\kappa + k\right)\right]}; \qquad \kappa = \frac{\sqrt{2m\left(E + V_0\right)}}{\hbar} \qquad (7.57)$$

As $a \to 0$, Eq. (7.54) has the solution obtained above for the square barrier, but the reflection coefficient (7.57) is smaller. For example, for $a = 1$ Å $V_0 = 10$ eV and $E = 0.1$ eV, one finds $R = 0.235$.

Chapter 8

Analytical Properties

The analytical properties of physical quantities enable us to study the general features of physical phenomena without the use of specific models. Generally, the functions that describe the physical phenomena are analytic functions of their arguments. The points at which a function is not analytic are called singular points. Such points include poles, where a function tends to infinity, points of discontinuity, where the function is not continuous, branch points, where one passes from one branch of a function to another and bifurcation points, where a change occurs in the number of possible solutions. The best-known example of bifurcation points are phase transitions, where the thermodynamic or kinetic parameters may have singularities. In Chapter 3, we considered the superfluidity transition which occurs even in an ideal system. This transition is an exception to the general rule that a phase transition is induced by a special form of interaction. The typical example is the ideal gas which exhibits no phase transitions, and the van-der-Waals gas, where the addition of interactions gives rise to a phase transition. We shall consider the analytical properties of some functions discussed in previous sections, such as the wave function and the scattering amplitude. The final part of this Chapter is devoted to symmetry.

8.1 Analytical properties of the wave function

The singularity of the wave function for different spherically symmetric potentials $V(r)$ follows from Eq. (7.20), where, for a non-singular potential function $V(r)$, the point $r = 0$ is the only singular point of the coefficients in this equation. According to Poincare's theorem, the solution of a linear differential equation can have singularities in the finite plane only at points where the coefficients appearing in the equation are singular. However, the presence of singularities of the function $\chi(r)$ depends on the form of the potential. If $V(r)$ is finite at the origin or tends to infinity at the origin less rapidly than r^{-2}, the function $\chi(r)$ is finite everywhere, being proportional to r^{l+1} for small r. For example, for the harmonic oscillator, the potential $V(r) = kr^2/2$ does not have singularities in the finite plane, and accordingly, the solution of the Schrödinger equation for the oscillator may have a singularity only at $r = \infty$. On the other hand, if $V(r)$ is negative near the origin and proportional to r^{-n} with $n \geq 2$, the radial part of the wave function $\Psi(r) = \chi(r)/r$ will have a singularity at $r = 0$, with the nature of this singularity depending on the detailed form of $V(r)$. This result is quite general. Singularities of the wave function can arise only from singularities of the potential. However, the converse is not true; singularities of the potential need not lead to a singularity of the wave function. As an example, the point $r = 0$ is a singular point of the coefficients of Eq. (7.20). Let the potential $V(r)$ for small r be smaller than $l(l+1)/r^2$. Then, Eq. (7.20) reduces to

$$-r^2 \frac{d^2\chi}{dr^2} + l(l+1)\chi = 0 \tag{8.1}$$

whose solutions are $\chi \approx r^{l+1}$ and $\chi \approx r^{-l}$, i.e., there exists a solution $\chi \approx r^{l+1}$ that remains finite at $r = 0$.

8.2   Analytical properties of the scattering amplitude

As we saw in Section 7, the analytic solution of Eq. (7.20) for the restricted potential $V(r)$ has the following form

$$\chi = A \sin(kr + \delta_l) \tag{8.2}$$

Equation (8.2) can be rewritten as

$$\chi = \frac{iA}{2} \exp(-i\delta) [\exp(-ikr) - S(k)\exp(ikr)] \tag{8.3}$$

This establishes the connection between the phase shift δ_l and the scattering matrix S, where $S(k) = \exp[2i\delta_l(k)]$ and $S(-k) = S^*(k)$. Since δ_l is a real function of k, the matrix $S(k)$ is unitary,

$$S^*(k)S(k) = 1 \tag{8.4}$$

One may average the asymptotic form of the wave function (7.1) over angles,

$$\langle \chi \rangle = \frac{1}{4\pi} \int \left[\exp(ikz) + f(\theta) \frac{\exp(ikr)}{r} \right] \sin\theta d\theta d\phi =$$

$$\frac{i}{2k} [\exp(-ikr) - (1 + 2ikf)\exp(ikr)] \tag{8.5}$$

where f is the averaged $f(\theta)$. Comparing (8.5) with (8.3) shows the connection between the S-matrix and the averaged scattering amplitude,

$$S(k) = 1 + 2ikf \qquad (8.6)$$

This formula allows one to rewrite the unitarity requirement (8.4) in terms of f,

$$|1 + 2ikf|^2 = 1 \qquad (8.7)$$

Since k is real, alternate forms of Eq. (8.7) are

$$\mathrm{Im}\, f = kf^2, \quad \mathrm{Im}\,\frac{1}{f} = \mathrm{Im}\,\frac{f^*}{f^2} = -k, \quad \mathrm{Re}\,\frac{1}{f} = g\left(k^2\right) \qquad (8.8)$$

where $g\left(k^2\right)$ is a real function of k^2,

$$f\left(k^2\right) = \left[g\left(k^2\right) - ik\right]^{-1} \qquad (8.9)$$

The quantity $g(0)$ can be expressed [37] in terms of the energy of the bound state in the potential well (if such a state exists).

Equation (8.8) determines the imaginary part of f^{-1}, while the real part depends on the form of the scattering potential $V(r)$. Equation (8.8) represents the special case considered in section 7.8, the optical theorem, which links the imaginary part of scattering cross-section through zero angle with the total cross-section. In general, the optical theorem establishes a link between the imaginary and real parts of the scattering amplitude, which determine the scattering of a particle and the dispersion relation, respectively.

8.3 Resonance scattering

Additional analytical properties of the $S-$matrix and the scattering amplitude $f(\theta)$ can be studied by extending the wave vector k to the imaginary axis, $k = i\kappa$. The energy $E = \hbar^2 k^2/2m$ then becomes negative, $E = -\hbar^2\kappa^2/2m$, describing a bound state. The wave function (8.3) will describe this bound state when terms in Eq. (8.3) containing the positive exponential ikr vanish. In other words, the function $S(\kappa)$ must vanish at points on the negative imaginary axis corresponding to the energies of such states. It then follows from Eqs. (8.6)-(8.9) that $S = 0$ leads to $g\left(k^2\right) = -ik$, and for each zero of $S(k)$ on the negative imaginary axis, there is a pole at the corresponding point on the positive imaginary axis.

Inserting (8.9) into (8.5) gives

$$\langle \chi \rangle = \frac{1}{2\kappa} \left[\exp(\kappa r) - \frac{g - \kappa}{g + \kappa} \exp(-\kappa r) \right] \tag{8.10}$$

The pole in $S(k)$ for $k = i\kappa = -ig(-\kappa^2)$ implies that the scattering sharply increases ("resonance scattering") when the energies of the incident particle and the bound states of the scattering potential have the same absolute value. Our results apply to the rapidly decreasing potential $V(r)$ and for s-wave scattering $(l = 0)$. The extension of these results can be found in [35], which includes the analysis of the analytical properties of the scattering amplitude at zero angle, $f(\theta = 0, E)$.

8.4   Symmetry

Symmetry plays a special role in quantum mechanics, yielding important general results without the need to perform any calculations. Before turning to the general approach, let us consider a few examples. The main properties of quantum systems are given by the Hamiltonian H, which depends on the set of variables ξ, $H = H(\xi)$. The symmetry of a system means that after performing some operation on the variables, $\xi \to -\xi$ or $\xi \to \xi + \delta\xi$ (reflection or shift in coordinates and/or in time, etc.), the Hamiltonian remains unchanged, $H(\xi + \delta\xi) = H(\xi)$. However, only $|\Psi|^2$ has physical meaning. Therefore, if some symmetry operation $\xi \to \xi + \delta\xi$ leads to the following change of the wave function, $\Psi(\xi + \delta\xi) = \exp(i\phi)\Psi(\xi)$ with real phase ϕ, the physical state of the system remains unchanged because $|\Psi(\xi)|^2 = |\Psi(\xi + \delta\xi)|^2$.

8.4.1   *Parity*

Consider a particle moving in a spherically symmetric field $V(\mathbf{r}) = V(-\mathbf{r})$. The Hamiltonian of the system then possesses the property $H(\mathbf{r}) = H(-\mathbf{r})$. Let the inversion operation $\mathbf{r} \to -\mathbf{r}$ be denoted by P, so that $P\Psi(\mathbf{r}) = \Psi(-\mathbf{r})$. The eigenvalues p of the operator P can easily be found by a double application of this operator, $P^2\Psi(\mathbf{r}) = p^2\Psi(\mathbf{r}) = \Psi(\mathbf{r})$ implying that $p = \pm 1$. Therefore, all states $\Psi(\mathbf{r})$ have a specific parity, $P\Psi(r) = +\Psi(\mathbf{r})$ or $P\Psi(\mathbf{r}) = -\Psi(\mathbf{r})$. The parity is a specific property of each given system. Indeed,

$$PH(\mathbf{r})\Psi(\mathbf{r}) = H(-\mathbf{r})\Psi(-\mathbf{r}) = H(\mathbf{r})P\Psi(\mathbf{r}) \tag{8.11}$$

or $PH = HP$. As we have seen in Section 1.3.3, the eigenvalues of all operators which commute with the Hamiltonian are conserved.

Parity is an internal property which does not exist in classical mechanics, and is unique to quantum objects. Many physical phenomena (electrodynamics, gravity, strong interactions in nuclear physics) are invariant with respect to the inversion of the coordinates. However, an exception is the weak interaction in nuclear physics to which the conservation of parity does not apply.

Another property of a particle moving in a spherically symmetric potential is the degeneracy of the quantum states. Degeneracy means that there are several states $\Psi_i(\mathbf{r})$ having the same energy. As we have seen in Section 5.6, the state of the electron in a spherically symmetric potential is characterized by three quantum numbers N, l and m, which define the energy, the angular momentum and the projection of the angular momentum on some (say, z) axis, respectively. In non-relativistic quantum mechanics, the energy depends only on the quantum number N while in the relativistic quantum mechanics, there appears an additional dependence on the quantum number l. Even in the latter case, the $(2l + 1)$ states with different quantum number m have the same energy. This degeneracy is easy to understand because in a spherically symmetric potential, all spatial axes are equivalent.

8.4.2 *Conservation laws*

The solution of the Schrödinger equation for a closed system (without external fields) can be written in the following form

$$\Psi(t + dt) = \exp\left(-\frac{iEdt}{\hbar}\right)\Psi(t) \qquad (8.12)$$

Equation (8.12) means that the translation of the wave function from time t to time $t + dt$ is described by the time translation operator $U_t = \exp(-iEdt/\hbar)$. Since all moments of time are equivalent for a closed system, the isotropy of time leads to the conservation law for energy. In an analogous way, using the series expansion

$$\Psi(r + dr) = \sum \frac{(dr\nabla_r)^k}{k!}\Psi(r) \qquad (8.13)$$

one can define the space translation operator U_r by

$$U_r = \exp(dr\nabla_r) \qquad (8.14)$$

which, analogously to Eq. (8.12), defines the conservation law of the radial momentum, described by the operator $-i\hbar\nabla_r$.

In both preceding examples, we introduced the translation in time and space operators U_t and U_r, which have the form

$$U = 1 - iRds \tag{8.15}$$

where $R = E/\hbar$, $ds = dt$ and $R = p/\hbar$, $ds = dr$ for infinitesimal translations in time and space, respectively. However, we know that rotation in space is connected with the conservation of angular momentum. Therefore, for $ds = d\phi$, we put $R = \mathbf{J}$, where $\mathbf{J}$ is the angular momentum operator. In the case of orbital angular momentum $\mathbf{J} = \mathbf{r} \times \mathbf{p}$, but in the case of the spin angular momentum, $\mathbf{J}$ has no classical analog.

8.4.3 *Degeneracy*

Generalizing the example of the degeneracy of the electron energy levels in a spherically symmetric potential, one can prove the following theorem. If some system has two conserved quantities, a and b, described by two non-commuting operators A and B, then the energy levels of such a system are degenerate. Indeed, the conservation of a and b means that their operators commute with the Hamiltonian H, $AH = HA$ and $BH = HB$. Let ϕ_n be the common eigenfunction of the commuting operators B and H

$$B\phi_n = b_n\phi_n; \qquad H\phi_n = E_n\phi_n \tag{8.16}$$

Applying operator A to the second equation gives

$$AH\phi_n = HA\phi_n = E_nA\phi_n \tag{8.17}$$

which shows that $A\phi_n$ is also an eigenfunction of the Hamiltonian H. But this function cannot be an eigenfunction of operator A because $A\phi_n$ is different from ϕ_n. Therefore, there is at least double degeneracy of energy level E_n, given by the two eigenfunctions ϕ_n and $A\phi_n$. In the case of an electron in a spherically symmetric potential, these two operators might be the two non-commuting operators of the projections of the angular momentum on the x- and y-axes. Another example is two symmetry operations, space inversion and translation, defined by the operators U_i and U_r, respectively. Assume that both these operators commute with the Hamiltonian H of a system,

$$HU_i = U_iH; \qquad HU_r = U_rH \tag{8.18}$$

However, momentum and parity do not commute, $U_rU_i \neq U_iU_r$, i.e., momentum eigenstates do not have definite parity, and parity states do not have definite non-zero momentum. Equation (8.18) implies that both these states have the same energy. One can also find [38] other general properties of such a system without knowing anything more than is given for the Hamiltonian.

8.4.4 *Internal symmetries*

In addition to the continuous (translation in space and time) and discrete (reflection, rotation) symmetries considered above, there are also internal symmetries. These refer to the different types of elementary particles. The strong and weak nuclear forces are short-range, with the radius of force 10^{-15} m and 10^{-18} m. These forces are responsible for the binding of nuclei and for radioactivity, respectively. The discovery of the neutron raised the following question: why do neutrons and protons, which differ in electric charge, have such similar masses, $m_p = 0.9986 \, m_n$, and why is the interaction between a proton and a neutron the same as the interaction between two protons or two neutrons? Heisenberg suggested in 1932 that the proton and the neutron are two states of the same particle ("nucleon"), having different quantum numbers, called isospin (or isotopic spin or isobaric spin). The isospin quantum number is conserved in strong interactions. Similar to the spin of a particle, the magnitude of the isospin has the value 1/2, with two projections $\pm 1/2$ corresponding to the proton and the neutron. Proton-neutron symmetry corresponds to the conservation law for isospin and its projections in all reactions involving strong interactions. The idea of isospin opened up new fields of internal symmetries and the corresponding conservation laws, such as the conservation laws for the baryon number and for the lepton number. The concept of isospin symmetry was later broadened to the larger symmetry groups of different types of quarks with appropriate conservation laws of quark flavor. However, these details of the theory of elementary particles are beyond the scope of our analysis.

8.4.5 *Complex conjugation and time reversal*

The Schrödinger equation contains the imaginary quantity i. Therefore, the solution of this equation $\Psi(\mathbf{r}, t)$ is complex. Although only $|\Psi|^2 = \Psi \Psi^*$ has physical meaning, it is interesting and important to find the correspondence between Ψ and Ψ^*. The complex conjugate equation of the Schrödinger equation has the following form

$$H\Psi^* = -i\hbar\frac{\partial\Psi^*}{\partial t} = i\hbar\frac{\partial\Psi^*}{\partial(-t)}, \qquad (8.19)$$

Thus, $\Psi^*(-t)$ is the solution of the Schrödinger equation which describes the "opposite-in-time" motion. These two functions, Ψ and Ψ^*, can differ in phase, which does not change $|\Psi|^2$. If the Hamiltonian is real, from the two equations,

$$H\Psi = E\Psi \text{ and } H\Psi^* = E\Psi^*, \qquad (8.20)$$

one can form the linear combinations $\Psi+\Psi^*$ and $i(\Psi - \Psi^*)$, which are both real, and contain the full set of real eigenfunctions. The situation is more complicated if a system has an internal degree of freedom, such as spin. Then, the transformation of the state includes the "orbital" change of the wave function considered above, and also the change of the "intrinsic" (spin) part of the wave function. The parity of a state, which is a product of the orbital and intrinsic parity, can be determined experimentally in processes for which parity is conserved. However, the intrinsic parity of an electron cannot be measured since electrons are only produced in pairs. Another peculiarity of particles with half-integral spin relates to their rotation, since the phase of their rotation state is not single-valued. Indeed, the rotation of state Ψ through an angle ϕ is determined by the total angular momentum J as $\Psi \to \exp(iJ\phi)\Psi$. The state of a system with half-integral eigenvalues is determined by the eigenvalues of the operator J. Hence, the rotation through the angle $\phi = 2\pi$ is not the identity operation, but leads to a change in sign of the function Ψ. This feature is not manifest in any measured quantity since particles of half-integral spin are always created or absorbed in pairs and all calculations contain even numbers of wave functions.

8.4.6 *Gauge transformation*

The Hamiltonian for a free particle in a magnetic field is the same as that of a free particle without a magnetic field, if one replaces the momentum **p** by $\mathbf{p} - e\mathbf{A}/c$, where **A** is the vector potential of the magnetic field. Since $\mathbf{B} = \nabla \times \mathbf{A}$, there is no magnetic field for constant **A**, but the Hamiltonian still contains the additional **A**-term. The solution of the Schrödinger equation is a plane wave, which is the same as for a free particle, $\Psi_\mathbf{k} = \exp(i\mathbf{kr})$, but the eigenvalues (relation between energy E and wave vector **k**) are different,

$$H\Psi_\mathbf{k} = E\Psi_\mathbf{k} = \frac{(\hbar\mathbf{k} - e\mathbf{A}/c)^2}{2m}\Psi_\mathbf{k} \qquad (8.21)$$

However, if one expresses the energy not as a function of the wave vector, but as a function of the velocity, whose operator has the form

$$v_x = \frac{dx}{dt} = \frac{i}{\hbar}(Hx - xH) = \frac{p_x - eA_x/c}{m}, \tag{8.22}$$

we return to the usual relation $E = \frac{1}{2}mv^2$. The way to avoid this ambiguity is by a gauge transformation of the wave function $\Psi \rightarrow \exp(iS)\Psi$ where $S = -e\mathbf{A}\mathbf{r}/\hbar c$, which transforms the state $\exp(i\mathbf{k}\mathbf{r})$ into $\exp(i\mathbf{k}^{\bullet}\mathbf{r})$ where $\mathbf{k}^{\bullet} = \mathbf{k} - e\mathbf{A}/\hbar c = m\mathbf{v}/\hbar$. This gauge transformation transforms the Hamiltonian to the usual form of a free particle

$$\exp(iS)\,H\exp(-iS) = \frac{p^2}{2m} \tag{8.23}$$

The procedure described above is a special case of a gauge transformation which may change the form of the Hamiltonian and of the wave function, but remains invariant to all physically measurable quantities.

Chapter 9

Paradoxes in Quantum Mechanics

In the previous Chapter we considered the analytical properties of physical variables, which follow from the general properties of space and time and analysis of the basis of quantum mechanics. Such an analysis not only allows one to obtain important results without performing any calculations, but also provides useful understanding of quantum mechanics. The analysis of paradoxes serves the same purpose. In 1911 Rutherford studied the scattering of alpha particles by atoms of gold and found that the atom has a positively charged nucleus. This result seems to be "paradoxical" because according to classical electrodynamics, the positive and negative charges must cancel due to the attractive forces. As we know, this classical paradox gave rise to quantum mechanics. The strange "paradoxical" behavior of quantum objects stimulated theories such as "hidden variables".

9.1 Schrödinger's cat

The next two paradoxes are formulated through thought experiments, interpretations of which lead to a seeming absurdity. As we have seen, all states in quantum mechanics can be divided into two groups: pure states, having a certain value of a given physical quantity and described by the eigenfunction of the associated quantum operator, and mixed states, where a physical quantity does not have a definite value and the system is described by the sum of appropriate eigenfunctions. However, upon performing the appropriate measurement, one always obtains a definite value of the physical quantity. In other words, our measurement transforms the mixed state into a pure state. This raises the question of what is the state of a system before the measurement. In dramatic form, this question leads to the paradox of Schrödinger's cat.

In this thought experiment, one imagines a cat sitting in a box with a radioactive source that has a 50 − 50 chance of emitting a poison that will kill the cat. If it does not emit the poison, the cat lives. We do not know, of course, whether the cat is alive or dead until the box is opened. There is no problem with the classical explanation of this situation. Indeed, the cat is either dead or alive depending on whether the radioactive source emitted a poison or not. We open the box in order to know what already happened previously in the closed box. However, the situation becomes more complicated if we treat it in a quantum manner. According to quantum mechanics, the cat in a box is described by the superposition of two states. One state corresponds to the cat being alive and the other state to the cat being dead. The probability of these two states is 50 − 50. Since the probabilities of these two states are equal, we cannot say whether the cat is alive or dead. Strictly speaking, we have to say that the poor cat is described by the sum of these two states, being 50% alive and 50% dead! However, when we make an observation by opening the box, the superposition of the two states suddenly collapses into an alive cat or a dead cat. Moreover, the transformation of the two states in one single state occurs instantly upon opening the box, in spite of the fact that the radioactive source had a 50 − 50 chance of killing the cat at any time during the cat's captivity in the box. This thought experiment contradicts common sense.

9.2 The Einstein-Podolsky-Rosen (EPR) paradox

The strange relation between classical and quantum mechanics connected with the measurement procedure can also be illustrated by the EPR paradox.

In their article, bearing the polemic title, "Can the quantum mechanical description of physical reality be considered complete?", EPR set out to prove the incompleteness of the quantum mechanical description of physical reality. EPR considered a system of two particles, each having spin angular momentum of magnitude $1/2$ (in appropriate units). Prepare the system with the spins pointing in opposite directions, say, in the z-direction Thus, the total spin of the system equals zero, with one particle having a z-component of spin $+1/2$, whereas the other particle has a z-component of spin $-1/2$.

One then separates these two particles in a way that does not affect their angular momentum, which is easily done. Say that one particle has been moved to the North Pole while the other particle is at the South Pole. One now measures the z-component of the spin of one particle, say, the North Pole particle, and obtains $-1/2$. What can one say about the measured value of the z-component of the spin of the South Pole particle?

According to classical physics, the answer is clear. The South Pole particle must have $+1/2$ for the z-component of spin because the sum of the z-components of both spins must equal zero.

Now let us consider this same question according to quantum mechanics. The answer is exactly the same! Since the operator of any component of angular momentum (say, the z-component) commutes with the operator for the total angular momentum, the measured values of z-component of the spin for the two particles remain $+1/2$ and $-1/2$ at all subsequent times, with 100% probability, exactly as in classical mechanics.

The situation is radically changed if one measures the component of the angular momentum in a different direction, say, at an angle α from the z-axis. First consider classical physics. The component of the angular momentum of the two particles in the α-direction will be $+(1/2)\cos\alpha$ and $-(1/2)\cos\alpha$, giving a sum of zero, as required. Thus, if the measured value of the North Pole particle is $+(1/2)\cos\alpha$, the measured value of the South Pole particle must be $-(1/2)\cos\alpha$, in order to maintain a zero value for their sum.

According to quantum mechanics, the situation is very different. Quantum mechanics states that there are only two possible results for the measurement of any component of angular momentum: $+1/2$ and $-1/2$, regardless of which direction the measurement is carried out. Therefore, also in the α-direction, the measurement will yield either $+1/2$ or $-1/2$. However, the probability of obtaining either of these two values is not 100%. Quantum mechanics provides the procedure for calculating the probability for getting each of these two values ($+1/2$ and $-1/2$) as the result of a measurement. But, before the measurement is performed, it is in principle impossible to know what the result of the measurement will be.

Now comes the paradox. If we measure the angular momentum of the North Pole particle in the α-direction, and obtain, say $+1/2$, then the value of the angular momentum of the South Pole particle in the α-direction must be $-1/2$, in order to make the total angular momentum equal to zero. However, before the measurement of the North Pole particle, it was not yet determined what the result of the measurement would be. Only after the measurement is it determined that the result is $+1/2$. Nevertheless, the South Pole particle is somehow influenced by the measured value of the

North Pole particle and will have value $-1/2$! Somehow, the message that the North Pole particle has a value of $+1/2$ is immediately transmitted all the way to the South Pole. Einstein called this seemingly absurd result, "spooky action at a distance".

It should be emphasized that the strange feature of this quantum prediction is not due to its violation of relativity theory, which states that no information can be transmitted faster than the speed of light. The paradox here is that we seem to have an example of action at a distance, which is impossible. One need only imagine that two particles to be at the opposite ends of the universe to see the magnitude of the paradox.

EPR claimed that this conclusion is so false that one has to complete the quantum mechanical theory by introducing some hidden variables. The "hidden" variables are the microscopic characteristics of quantum objects, which cannot be observed and measured because of restrictions of existing devices. By contrast, the uncertainty principle of quantum mechanics says that some pairs of variables (coordinate and momentum, or energy and time) simple cannot exist simultaneously. This is the property of the real world, and has nothing to do with measurements or the type of devices.

9.3   Hidden variables and Bell's inequality

The EPR paradox discussed above casts doubt on the basic principle of quantum mechanics. Either the EPR experiment violates the principle of the finite propagation speed of physical phenomena, or it is incomplete by not taking into account some hidden physical variables. The latter can be exemplified by the following example from classical mechanics. When a roulette wheel in a casino stops on some winning number, the happy winner reaps the harvest of this "random" success. However, it is obvious that if one had detailed knowledge of the "hidden" parameters (initial speed of the wheel, friction, etc.), one could calculate in advance the winning number. The question is whether quantum mechanics is statistical in its essence, or there are some additional ("hidden") variables, different for the two particles in the EPR experiment. If such hidden variables exist, they can restore determinism. Without these variables, quantum mechanics seems to be incomplete. In a general way, the difference between classical mechanics and quantum mechanics is that in classical mechanics, a system is characterized by physical parameters that can be known before the measurement, which only reveals these parameters. By contrast, in quantum mechanics these parameters are created during the measurement.

The history of physics supports the idea of hidden variables. Before knowledge of the atomic structure of matter, scientists tended to explain new experimental results by various modifications of existing theories. It was the discovery of atoms, "hidden" (before their discovery) variables that furnished insight into the nature of the phenomena studied. Another example is Newton's laws of mechanics, which explain many dynamic problems. These laws turned out to be "incomplete" for the explanation of some cosmological problems or of motion with velocity close to that of light, where one needs to add "hidden" knowledge from the theory of relativity.

EPR argued that the strange results of their thought experiment also proves that quantum mechanics is incomplete, and the full explanation of this experiment can only be achieved by introducing some "hidden" variables, which will enable one to return to a deterministic description of Nature. Einstein disliked the probabilistic nature of reality, noting that "God does not play dice". In some sense, the usual (Copenhagen) interpretation of quantum mechanics bypasses the question of determinism or randomness in Nature by dividing its existence into before the measurement, when each object might be in many different positions, and after the measurement, by which the experimentalist "creates" its position. Such an approach produces deep philosophical problems. If "hidden" variables exist, one can avoid these problems and return to the deterministic approach of Newtonian mechanics.

In 1964, John Bell derived an inequality which could be used to test for the existence of hidden variables [41]. If this inequality is not satisfied, then hidden variables do not exist. We here present a brief version of Bell's inequality; a more detailed description is given in [42]].

Consider an observer who measures the three spin components of an ensemble of spin $1/2$ particles along three axes (A, B, C) that may point in any three directions. In each direction, there are only two possible results for the measurement of the spin component: $+1/2$ or $-1/2$. Let $F(A_+, B_-, C_+)$ denote the fraction of the particles in the ensemble having spin components $+1/2$ in direction A, $-1/2$ in direction B, and $+\frac{1}{2}$ in direction C, in an obvious notation. Similarly, let $F(A_-, C_+)$ denote the fraction of particles in the ensemble having spin component $-\frac{1}{2}$ in direction A and $+\frac{1}{2}$ in direction C, regardless of the value of the spin component in direction B.

The important point to note is that the functions $F(A_+, B_-, C_+)$ and $F(A_-, C_+)$ do not exist according to quantum mechanics. A definite value of the spin component (pure quantum state) can exist in only one direction. After choosing that direction, one has a mixed quantum state in any other direction, meaning that the value of the spin component in any other direction is determined only after a measurement. For a mixed state in quantum mechanics, there is no meaning to the question: What was the value of the spin component before the measurement? Before the measurement, the value of the spin component could not be known even in principle.

In complete contrast to quantum mechanics, in classical physics with hidden variables, the values of the spin components always exist in all directions and, hence, the F-functions do exist. A measurement only reveals what already exists. There may be technical problems in measuring the spin component, but there is no problem in principle. We shall continue the discussion assuming the existence of hidden variables, which implies that there is meaning to the F-functions.

The definition of $F(A_+, C_-)$ is

$$F(A_+, C_-) = F(A_+, B_+, C_-) + F(A_+, B_-, C_-) \qquad (9.1)$$

Equation (9.1) implies

$$F(A_+, C_-) > F(A_+, B_+, C_-) \qquad (9.2)$$

We may ignore the possibility of an equality sign if $F(A_+, B_-, C_-)$ happens to vanish.

An analogous argument gives

$$F(B_+, C_-) > F(A_-, B_+, C_-) \qquad (9.3)$$

Adding (9.2) and (9.3) gives

$$F(A_+, C_-) + F(B_+, C_-) > F(A_+, B_+, C_-) \qquad (9.4)$$
$$+ F(A_-, B_+, C_-) = F(B_+, C_-)$$

where the last equality follows from the definition of $F(B_+, C_-)$. Therefore, (9.4) states the important result that

$$F(B_+, C_-) < F(A_+, C_-) + F(B_+, C_-) \qquad (9.5)$$

This is Bell's inequality, a result that apparently can never be tested experimentally. One cannot measure the F functions because it is impossible to measure simultaneously the spin components in two different directions. The measurement of the spin component in, say, direction B, alters the values of the spin component in directions A and C. Therefore, all three F-functions in (9.5) seem unknowable, and Bell's inequality seems completely useless.

However, John Bell proposed the following ingenious solution to this problem of how to measure the F-functions. Let us perform an EPR experiment. That is, one prepares a pair of spin $-1/2$ particles whose spins are oriented anti-parallel, implying total spin zero. Therefore, the sum of the spin components of these two particles is also zero in any direction. Define $G_1(B_+)$ as the probability of particle 1 having $+1/2$ for the spin component in direction B, with an analogous definition for $G_2(B_-)$. But, one knows that $G_2(B_-) = G_1(B_+)$, since the sum of the spin components of the two particles in the same direction B was arranged to be zero. Whenever the measured value of the spin component of particle 2 in direction B is $-1/2$, one knows for certain, without performing any measurement on particle 1, that the spin component of particle 1 in direction B is $+1/2$. Therefore – and here is the point – one can determine the value of the spin component of particle 1 in direction B by performing a measurement on particle 2, without disturbing particle 1 at all! Thus, one can measure the F-functions, and Bell 's inequality can be experimentally tested.

To summarize, one determines $F(B_+, C_+)$ for particle 1 by simultaneously measuring both the probability of obtaining $1/2$ for the spin component of particle 2 in direction B (knowing that particle 1 must have the opposite spin component) and also measuring the probability of obtaining $+1/2$ for the spin component for particle 1 in direction C. This is perfectly feasible since one never measures the spin component of the same particle in two different directions.

Many experiments have been carried out to test Bell's inequality. An analysis of 13 such experiments [43] shows that Bell's inequality is violated and, therefore, the EPR assumption of hidden variables is incorrect. Moreover, the experimental results are in exact agreement with the predictions of quantum mechanics. This subject continues to be investigated with different modifications of the EPR experiment and new versions of Bell's inequality.

Bibliography

[1] S. Flugge, *Practical quantum mechanics* [Springer-Verlag, 1973].

[2] A. A. Kamal, *1000 solved problems in modern physics: an exercise book* [Springer, 2010].

[3] J. R. Boccio, *Introduction to quantum field theory*, e-book http://chaos. swarthmore.edu/courses/Physics093_2009/

[4] F. Constantinescu and E. Magyari, *Problems in quantum mechanics* [Pergamon, 1971].

[5] J.-L. Basdevant and J. Dalibard, *The quantum mechanics solver* [Springer, 2005].

[6] G. L. Squiers, *Problems in quantum mechanics with solutions* [Cambridge, 1995].

[7] B. M. Galickii, B. M. Karnakov, and V. U. Kogan, *Problems in quantum mechanics* (in Russian) [Moscow, 1981].

[8] I. I. Goldman and V. D. Krivchenkov, *Problems in quantum mechanics* [Dover, 2006].

[9] P. Satsidis, *Solutions to problems in Sakurai's quantum mechanics,* [Addison-Wesley, 1985].

[10] I. S. Gradstein and I. M. Ryzhik, *Tables of integrals, series and products* [Academic, 1971].

[11] J. G. Bednorz and K. A. Müller, Z. Phys. B **64**, 189 (1986).

[12] Y. Kamihara, H. Hiramatsu, M. Hiraro, R. Karamura, H. Yanagi, T. Kamiya, and H. Hosono, J. Am. Chem. Soc. **128**, 10012 (2006).

[13] J. A. Wilson, J. Phys.: Condens. Matter **22**, 203201 (2010); P. M. Aswathy, J. B. Anooja, P. M. Sarun, and U. Syamaprased, Supercond. Sci. Technol. **23**, 073001 (2010).

[14] R. Mitsuhashi, Y. Suzuki, Y. Yamanari, H. Mitamura, T. Kambe, N. Ikeda, H. Okamoto, A. Fujiwara, M. Yamaji, N. Kawasaki, Y. Maniwa, and Y. Kubozono, Nature **464**, 76 (2010).

[15] F. Steglich, J. Aarts, C. D. Bredl, W. Lieke, D. Meschede, W. Franz, and J. Schifer, Phys. Rev. Lett. **43**, 1892 (1979).

[16] Y. Maeno, H. Hashimoto, K. Yoshida, S. Nishizaki, T. Fujita, J. G. Bednorz, and F. Lichtenberg, Nature **372**, 532 (1994).

[17] J. L. Tallon, C. Bernard, and J. W. Loram, J. Low Temp. Phys. **117**, 823 (1999).

[18] J. Nagamateu, N. Nakagawa, T. Miranaka, Y. Zenitani, and J. Akimutsu, Nature **410**, 63 (2001).

[19] M. Jones and R. Marsh, J. Am. Chem. Soc. **76**, 1434 (1954).

[20] P. C. Canfield, D. K. Finnemore, S. L. Bud'ko, J. E. Ostenson, G. Lapertot, C. E. Cunningham, and C. Petrovic, Phys. Rev. Lett. **86**, 2423 (2001).

[21] J. E. Hirsch, Phys. Scr. **80**, 035702 (2009).

[22] D. I. Thouless, *The quantum mechanics of many-body systems* [Academic, 1961].

[23] R. D. Mattuck, *A guide to Feynman diagrams in the many-body problem* [McGraw-Hill, 1967].

[24] R. Gerritsma, G, Kirchmair, F. Zahringer, E. Solano. R. Blatt, and C. F. Roos, Nature **463**, 68 (2010).

[25] W. E. Lamb, Jr. and R. C. Retherford, Phys. Rev. **86**, 1014 (1952).

[26] M. I. Katsnelson, K. Novoselov, and A. Geim, Nature Physics **2**, 620 (2006).

[27] N. Dombey and A. Calogerasos, Phys. Reports **315**, 41 (1999); A. Calogerasos and N. Dombey, Contemp. Phys. **40**, 321 (1999).

[28] O. Klein, Z. Phys. **53**, 157 (1929).

[29] S. K. Lamoreaux, Phys. Rev. Lett. **78**, 5 (1997).

[30] G. Bressi, G. Carugno, R. Onofrio, and G. Ruoso, Phys. Rev. Lett. **88**, 041804 (2002).

[31] M. Borgad, G. Kimchitskaya, U. Mohideen, and V. Mostepanenko, *Advances in the Casimir effect* [Oxford Scientific Publications, 2009].

[32] I. E. Dzyaloshinskii, E. M. Lifshitz, and I. P. Pitaevskii, Adv. Phys. **10**, 165 (1961).

[33] www.electron9.phys.utk.edu/phys514/modules/module13/problems.

[34] D. J Vezzetti and M. M. Cahay, J. Phys. D **19**, L53 (1986).

[35] L. D. Landau and E. M. Lifshitz, *Quantum mechanics* [Elsevier, 1958].

[36] R. G. Newton, Amer. J. Phys. **44**, 639 (1976)

[37] A. B. Migdal, *Qualitative methods in quantum theory* [Westview, 2000].

[38] H. J. Lipkin, *Quantum mechanics* [North Holland, 1973].

[39] A. Einstein, B. Podolsky, and N. Rosen, Phys. Rev. **41**, 777 (1935).

[40] D. Bohm, *Quantum theory* [Dover, 1957].

[41] J. Bell, Physics **1**, 195 (1964); Rev. Mod. Phys. **38**, 437 (1966).

[42] A. Shimony, *Stanford encyclopedia of philosophy* [Stanford University, 2004].

[43] M. Readhead, *Incompleteness, nonlocality and realism* [Clarendom Press, 1987].

Index